Springer Earth System Sciences

Series editors

Philippe Blondel, Bath, UK
Eric Guilyardi, Paris, France
Jorge Rabassa, Ushuaia, Argentina
Clive Horwood, Chichester, UK

The Springer Earth System Sciences series focuses on interdisciplinary research linking the lithosphere (geosphere), atmosphere, biosphere, cryosphere, and hydrosphere that build the system earth. The series seeks to publish a broad portfolio of scientific books, aiming at researchers, students, and everyone interested in this extremely interdisciplinary field. It covers the entire research area of earth system sciences including, but not limited to, Earth System Modeling, Glaciology, Climatology, and Human-Environment/Earth interactions. Springer Earth System Sciences includes peer-reviewed monographs, edited volumes, textbooks, and conference proceedings.

More information about this series at http://www.springer.com/series/10178

Xiujun Wang · Zhitong Yu
Jiaping Wang · Juan Zhang
Editors

Carbon Cycle in the Changing Arid Land of China

Yanqi Basin and Bosten Lake

Editors
Xiujun Wang
College of Global Change and Earth System
  Science
Beijing Normal University
Beijing
China

Zhitong Yu
College of Global Change and Earth System
  Science
Beijing Normal University
Beijing
China

Jiaping Wang
Agricultural College
Shihezi University
Shihezi
China

Juan Zhang
College of Resources and Environment
Northeast Agricultural University
Harbin
China

ISSN 2197-9596 ISSN 2197-960X (electronic)
Springer Earth System Sciences
ISBN 978-981-10-7021-1 ISBN 978-981-10-7022-8 (eBook)
https://doi.org/10.1007/978-981-10-7022-8

Library of Congress Control Number: 2018932193

© Springer Nature Singapore Pte Ltd. 2018
This work is subject to copyright. All rights are reserved by the Publisher, whether the whole or part of the material is concerned, specifically the rights of translation, reprinting, reuse of illustrations, recitation, broadcasting, reproduction on microfilms or in any other physical way, and transmission or information storage and retrieval, electronic adaptation, computer software, or by similar or dissimilar methodology now known or hereafter developed.
The use of general descriptive names, registered names, trademarks, service marks, etc. in this publication does not imply, even in the absence of a specific statement, that such names are exempt from the relevant protective laws and regulations and therefore free for general use.
The publisher, the authors and the editors are safe to assume that the advice and information in this book are believed to be true and accurate at the date of publication. Neither the publisher nor the authors or the editors give a warranty, express or implied, with respect to the material contained herein or for any errors or omissions that may have been made. The publisher remains neutral with regard to jurisdictional claims in published maps and institutional affiliations.

Printed on acid-free paper

This Springer imprint is published by Springer Nature
The registered company is Springer Nature Singapore Pte Ltd.
The registered company address is: 152 Beach Road, #21-01/04 Gateway East, Singapore 189721, Singapore

Foreword

This book, "Carbon cycle in the changing arid land: Yanqi Basin and Bosten Lake," comes at a time when countries are seriously looking for ways to offset CO_2 emissions. Arid and semiarid regions make up about one-third of the Earth's land surface, yet these drylands are generally overlooked as places for carbon sequestration because (1) soil organic carbon is low compared to more humid soils and (2) soil inorganic carbon is widely considered to be an inert carbon reservoir rather than an active pool that can be manipulated by land management techniques.

This book provides examples suggesting that soil inorganic carbon is more dynamic and complicated than traditionally thought. In addition, readers of this book will learn about carbon in the interesting geological and climatic setting of northwest China. Perhaps its most important contribution, this book can stimulate carbon research in rangeland and irrigated agricultural settings in other arid land regions of the world.

Las Cruces, USA H. Curtis Monger, Ph.D.
Professor Emeritus, New Mexico State University

Contents

The Carbon Cycle in Yanqi Basin and Bosten Lake: Introduction 1
Xiujun Wang, Jiaping Wang, Zhitong Yu and Juan Zhang

Introduction of the Yanqi Basin and Bosten Lake 5
Changyan Tian, Lei Zhang and Shuai Zhao

Climate Change Over the Past 50 Years in the Yanqi Basin 19
Fengqing Jiang, Junyi Wang and Xiujun Wang

Characteristics of Soil Organic Matter and Carbon and Nitrogen
Contents in Crops/Plants: Land Use Impacts 41
Juan Zhang, Xiujun Wang, Jiaping Wang and Qingfeng Meng

Dynamics of Soil CO_2 and CO_2 Efflux in Arid Soil 55
Junyi Wang, Xiujun Wang, Jiaping Wang and Tongping Lu

Land Use Impacts on Soil Organic and Inorganic Carbon and
Their Isotopes in the Yanqi Basin 69
Jiaping Wang, Xiujun Wang and Juan Zhang

Distribution of Pedogenic Carbonate and Relationship with Soil
Organic Carbon in Yanqi Basin 89
Xiujun Wang, Jiaping Wang and Junyi Wang

Spatial Distribution of Organic Carbon in Surface Sediment
of Bosten Lake 103
Zhitong Yu, Xiujun Wang and Hang Fan

Temporal Variability of Carbon Burial and the Underlying
Mechanisms in Bosten Lake Since 1950 117
Zhitong Yu and Xiujun Wang

Carbon Sequestration in Arid Lands: A Mini Review 133
Xiujun Wang, Jiaping Wang, Huijin Shi and Yang Guo

Contributors

Hang Fan College of Global Change and Earth System Science, Beijing Normal University, Beijing, China

Yang Guo College of Global Change and Earth System Science, Beijing Normal University, Beijing, China

Fengqing Jiang State Key Laboratory of Desert and Oasis Ecology, Xinjiang Institute of Ecology and Geography, Chinese Academy of Sciences, Urumqi, Xinjiang, China

Tongping Lu College of Global Change and Earth System Science, Beijing Normal University, Beijing, China

Qingfeng Meng School of Resources and Environment, Northeast Agricultural University, Harbin, China

Huijin Shi College of Global Change and Earth System Science, Beijing Normal University, Beijing, China

Changyan Tian State Key Laboratory of Desert and Oasis Ecology, Xinjiang Institute of Ecology and Geography, Chinese Academy of Sciences, Urumqi, Xinjiang, China

Jiaping Wang College of Agriculture, Shihezi University, Shihezi, China

Junyi Wang College of Global Change and Earth System Science, Beijing Normal University, Beijing, China

Xiujun Wang College of Global Change and Earth System Science, Beijing Normal University, Beijing, China

Zhitong Yu College of Global Change and Earth System Science, Beijing Normal University, Beijing, China; Xinjiang Institute of Ecology and Geography, Chinese Academy of Sciences, Urumqi, China

Juan Zhang School of Resources and Environment, Northeast Agricultural University, Harbin, China

Lei Zhang State Key Laboratory of Desert and Oasis Ecology, Xinjiang Institute of Ecology and Geography, Chinese Academy of Sciences, Urumqi, Xinjiang, China

Shuai Zhao State Key Laboratory of Desert and Oasis Ecology, Xinjiang Institute of Ecology and Geography, Chinese Academy of Sciences, Urumqi, Xinjiang, China

The Carbon Cycle in Yanqi Basin and Bosten Lake: Introduction

Xiujun Wang, Jiaping Wang, Zhitong Yu and Juan Zhang

1 Introduction

The rate of carbon dioxide (CO_2) build-up in the atmosphere depends on the rate of fossil fuel combustion and the rate of CO_2 uptake by the ocean and the land. About half of the anthropogenic CO_2 has been absorbed by the land and the ocean, the so-called sinks for CO_2. The efficiency of the global CO_2 sinks has been observed to change on seasonal to interannual and longer timescales, due to a variety of mechanisms that are not fully understood. Many of the mechanisms are not yet quantitatively defined either at regional or global scales.

The global soil carbon pool is the third largest pool in the Earth system, thus plays an important role in the global carbon cycle and climate system. Soil carbon pool consists of two components, soil organic carbon (SOC) and inorganic carbon (SIC). Soil organic carbon, as a key index for soil fertility and a means for carbon sequestration, has gained recognition. In contrast, much less has been done to determine the magnitude and variability of SIC and to understand SIC dynamics although scientists pointed out its potential for carbon sequestration and climate mitigation (Eshel et al. 2007; Lal and Kimble 2000; Zheng et al. 2011). To date, there is a large discrepancy in the estimated global SIC pool, which ranges from <700 to >1700 Pg (see Eswaran et al. 2000).

The SOC pool is the predominant carbon pool in soils of humid and semi-humid regions, whereas SIC is the most common form of carbon in soils of arid and semiarid zone. More than 35% of Earth's land surface is characterized as either arid or semiarid, where SIC content is 2–10 times as high as SOC content (Scharpenseel et al. 2000).

X. Wang (✉) · Z. Yu
College of Global Change and Earth System Science, Beijing Normal University, Beijing 100875, China
e-mail: xwang@bnu.edu.cn

J. Wang
College of Agriculture, Shihezi University, Shihezi 832000, China

J. Zhang
School of Resources and Environment, Northeast Agricultural University, Harbin 150030, China

© Springer Nature Singapore Pte Ltd. 2018
X. Wang et al. (eds.), *Carbon Cycle in the Changing Arid Land of China*, Springer Earth System Sciences, https://doi.org/10.1007/978-981-10-7022-8_1

In this regard, accurately estimating SIC at all scales is essential to evaluate the role of soils in the global carbon cycle (Yang and Li 2011). Moreover, attempts to decrease the atmospheric CO_2 concentration require better understanding of transformation of both SOC and SIC and regulating processes in different soils (Mikhailova and Post 2006; Monger and Gallegos 2000; Zheng et al. 2011).

2 Studies in Yanqi Basin

The arid and semiarid regions have experienced significant climate changes and human activity over the past decades. Climate change (e.g., warming) may have large impacts on the carbon cycle in desert ecosystem (Arnone et al. 2008; Verburg et al. 2005). There has been a significant increase in both temperature (Chen et al. 2009; Liu et al. 2005) and precipitation (Wang and Zhou 2005) in the northwest China over the past 50 years, with implications for the terrestrial ecosystem and carbon cycle (Liu et al. 2009). In addition, there is an increasing trend in growing season for the entire Xinjiang (Jiang et al. 2011). There is also evidence of land use and climate change impacts on SOC and SIC in the northwest China (Wu et al. 2003, 2009). However, there have been limited studies that apply an integrated approach to assess the impacts of climate change and land use changes on the carbon cycling at a regional scale in arid lands.

The news of "Have desert researchers discovered a hidden loop in the carbon cycle" (Stone 2008) prompted a focused effort to better understand the carbon cycle in the atmosphere–plant–soil systems in the arid land of northwest China. A few studies have yielded collections of a variety of datasets, including approximately 60 years of climate and hydrological data, nearly 40 profiles of SOC and SIC under various land uses, and seasonal variations of CO_2 efflux and soil CO_2 concentration at different depths in the Yanqi Basin. In addition, a study was carried out to investigate the spatial and temporal variations of OC and IC burials in Bosten Lake. These datasets provide basic information of key hydrological and biogeochemical processes in association with the carbon sequestration in the arid region. On the other hand, these data can also be used for carbon model calibration and validation for the arid region.

This book reports the studies conducted in the Yanqi Basin and Bosten Lake, which aim to better understand the carbon cycle in the arid lands. The first two chapters provide basic information for the region and motivations for the studies. Chapter 3 reports the analyses of climate variables (temperature, precipitation and runoff) over the period of 1960–2014. Chapters 4–9 are the outcomes of core studies on the carbon cycle, which include the distributions of carbon and nitrogen under different vegetation/crops (Chap. 4), the dynamics of soil CO_2 concentration and surface CO_2 efflux in agricultural soil (Chap. 5), the dynamics of SOC and SIC and their isotopes under various land uses (Chap. 6), pedogenic carbonate and its relationship with SOC (Chap. 7), and the spatial and temporal variations in carbon burial in the Bosten Lake (Chaps. 8 and 9). The last chapter is a mini review, which summarizes the main findings from relevant studies, and also discusses the implications and future directions.

3 Significance of These Studies

A comprehensive evaluation of both SOC and SIC stocks in soils/sediments is important to achieve a better understanding of the carbon cycle in the coupled atmosphere–biosphere–pedosphere system. The carbon cycle in the arid and semiarid lands has to deal with the dynamics of inorganic carbon. To date, most studies of the terrestrial carbon cycle have largely focused on SOC pool and atmosphere–land CO_2 fluxes, and our understanding is limited in terms of variability and underlying mechanisms. Apparently, we need data collections of various forms of carbon (e.g., SOC and SIC contents, soil CO_2 concentration and surface CO_2 efflux, and OC and IC in sediments) to understand the variability of carbon fluxes in the atmosphere–biosphere–pedosphere systems in the arid lands.

In summary, the outcomes of these studies will help answer the following important scientific questions: (1) **where does the carbon go if a significant amount of CO_2 is absorbed by surface soils in arid regions?** (2) **how do environmental changes impact the carbon cycle in the arid and semiarid lands?** Answers to these questions will provide insights of the carbon sinks/sources and their variability in the arid and semiarid regions, but also improve our understanding of the carbon cycle at regional to global scales.

References

Arnone JA, Verburg PSJ, Johnson DW, Larsen JD, Jasoni RL, Lucchesi AJ, Batts CM, von Nagy C, Coulombe WG, Schorran DE, Buck PE, Braswell BH, Coleman JS, Sherry RA, Wallace LL, Luo YQ, Schimel DS (2008) Prolonged suppression of ecosystem carbon dioxide uptake after an anomalously warm year. Nature 455:383–386. https://doi.org/10.1038/nature07296

Chen YN, Xu CC, Hao XM, Li WH, Chen YP, Zhu CG, Ye ZX (2009) Fifty-year climate change and its effect on annual runoff in the Tarim River Basin, China. Quatern Int 208:53–61. https://doi.org/10.1016/j.quaint.2008.11.011

Eshel G, Fine P, Singer MJ (2007) Total soil carbon and water quality: an implication for carbon sequestration. Soil Sci Soc Am J 71:397–405. https://doi.org/10.2136/sssaj2006.0061

Eswaran H, Reich PF, Kimble JM, Beinroth FH, Padmanabhan E, Moncharoen P (2000) Global carbon stocks. In: Lal R, Kimble JM, Eswaran H, Stewart BA (eds) Global climate change and pedogenic carbonates. Lewis Publishers, Boca Raton

Jiang FQ, Hu RJ, Zhang YW, Li XM, Tong L (2011) Variations and trends of onset, cessation and length of climatic growing season over Xinjiang, NW China. Theor Appl Climatol 1–10. https://doi.org/10.1007/s00704-011-0445-5

Lal R, Kimble JM (2000) Pedogenic carbonate and the global carbon cycle. In: Lal R, Kimble JM, Eswaran H, Stewart BA (eds) Global climate change and pedogenic carbonates. CRC Press, Boca Raton

Liu DX, Dong AX, Deng ZY (2005) Impact of climate warming on agriculture in Northwest China. J Nat Resour 80:379–393. https://doi.org/10.1007/s10584-006-9121-7

Liu WX, Zhang Z, Wan SQ (2009) Predominant role of water in regulating soil and microbial respiration and their responses to climate change in a semiarid grassland. Glob Change Biol 15:184–195

Mikhailova EA, Post CJ (2006) Effects of land use on soil inorganic carbon stocks in the Russian Chernozem. J Environ Qual 35:1384–1388. https://doi.org/10.2134/jeq2005.0151

Monger HC, Gallegos RA (2000) Biotic and abiotic processes and rates of pedogenic carbonate accumulation in the southwestern United States-Relationship to atmospheric CO_2 sequestration. In: Lal R, Kimble JM, Eswaran H, Stewart BA (eds) Global climate change and pedogenic carbonates. CRC Press, Boca Raton

Scharpenseel HW, Mtimet A, Freytag J (2000) Soil inorganic carbon and global change. In: Lal R, Kimble JM, Eswaran H, Stewart BA (eds) Global climate change and pedogenic carbonates. CRC Press, Boca Raton

Stone R (2008) Have desert researchers discovered a hidden loop in the carbon cycle? Science 320:1409–1410

Verburg PSJ, Larsen J, Johnson DW, Schorran DE, Arnone JA (2005) Impacts of an anomalously warm year on soil CO_2 efflux in experimentally manipulated tallgrass prairie ecosystems. Glob Change Biol 11:1720–1732. https://doi.org/10.1111/j.1365-2486.2005.001032.x

Wang Y, Zhou L (2005) Observed trends in extreme precipitation events in China during 1961–2001 and the associated changes in large-scale circulation. Geophys Res Lett 32:L09707

Wu H, Guo Z, Peng C (2003) Distribution and storage of soil organic carbon in China. Global Biogeochem Cycles 17:1048. https://doi.org/10.1029/2001gb001844

Wu HB, Guo ZT, Gao Q, Peng CH (2009) Distribution of soil inorganic carbon storage and its changes due to agricultural land use activity in China. Agr Ecosyst Environ 129:413–421. https://doi.org/10.1016/j.agee.2008.10.020

Yang LF, Li GT (2011) Advances in research of soil inorganic carbon. Chin J Soil Sci 42:986–990 (in Chinese)

Zheng J, Cheng K, Pan G, Pete S, Li L, Zhang X, Zheng J, Han X, Du Y (2011) Perspectives on studies on soil carbon stocks and the carbon sequestration potential of China. Chin Sci Bull 56:3748–3758. https://doi.org/10.1007/s11434-011-4693-7

Introduction of the Yanqi Basin and Bosten Lake

Changyan Tian, Lei Zhang and Shuai Zhao

Abstract The Yanqi Basin is formed on the basement of the Kuruktag fold belt and the southern Tianshan fold belt and located in the inland region of the Central Asia and in the transition zone between the Junggar Basin and the Tarim Basin. The Yanqi Basin is a typical arid region with extremely low precipitation (<100 mm/year) but strong evaporation (>2000 mm/year). The main soil types are brown desert soil and alluvial soil, which were developed from limestone parent materials. Land use types/coverages include cropland, shrubland, and desert land. The basin has access to water resources from the Kaidu River and underground waters, which are mainly from melting snows in the Tianshan Mountain. Bosten Lake is the largest freshwater lake in Xinjiang, which is the final converging place for the surface water and groundwater in the Yanqi Basin. Its main inflow water is from the Kaidu River on the west, and outflow is from the Kongque River on the southwest.

1 Geographical Position

Yanqi Basin (85° 50′–87° 50′E, 41° 40′–42° 20′N) is located in the region of Bayingolin Mongol Autonomous Prefecture, Xinjiang, Northwest China (Fig. 1), and this basin belongs to the Kaidu River–Kongque River watershed. It is a Mesozoic rift basin between the main ridge and the branch ridge of the Tianshan Mountain; thus, it is also a kind of semi-enclosed intermontane basin. It lies at an altitude of 1048–1160 m, spanning an area of 5600 km^2 (Kou et al. 2008).

C. Tian (✉) · L. Zhang · S. Zhao
State Key Laboratory of Desert and Oasis Ecology, Xinjiang Institute of Ecology and Geography, Chinese Academy of Sciences, Urumqi 830011, Xinjiang, China
e-mail: tianchy@ms.xjb.ac.cn

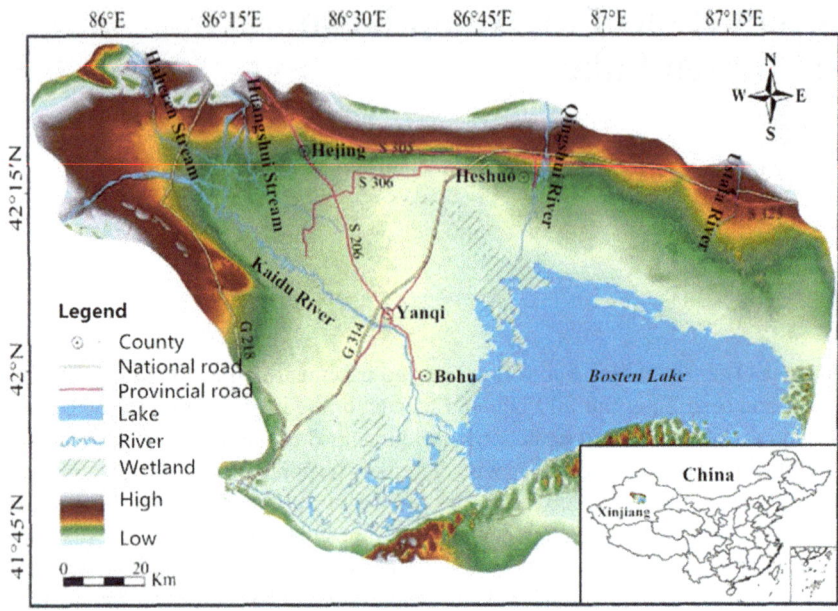

Fig. 1 Geographical position of Yanqi Basin (after Mamat Zulpiya 2014)

2 Geologic Landforms

2.1 Geology

The Yanqi Basin is known as a Mesozoic–Cenozoic basin, which is formed on the basement of the Kuruktag fold belt and the southern Tianshan fold belt. Its southern boundary is the Xingeer fault, borders Kuluketage mountain; its northern boundary is the Sangshuyuanzi fault, borders Saaerming mountain; its western boundary is the Tiemenguan fault, borders Huola mountain; its eastern boundary is the Kezile mountain. According to the distribution of Mesozoic and Cenozoic strata in the basin, its geometric structure can be divided into three tectonic units: Bohu depression in the south, Yanqi uplift in the middle, and Hejing depression in the north.

It has been found that the main sedimentary layers were formed in the middle-lower Jurassic Series, Neogene System, Paleogene System, and Quaternary System. The layers from the middle and middle-low Jurassic System consist of several coal seams and dull mudstone; the layers from the Neogene System consist of thick brick-red mudstone and argillaceous siltstone, which exist as stable areal cap rocks.

In the plain area of central basin, sediments were mainly formed in the Quaternary System and classified as diluvial deposit, alluvial–diluvial deposit, alluvial deposit, and/or alluvial–lacustrine deposit. The lithologic characters of diluvial plain and alluvial–diluvial plain exhibit a single thick layer structure in which only cobblestone exists or a double-layer structure in which a thin sandy loam soil layer and a thick cobblestone layer exist. The lithologic character of alluvial plain and alluvial–lacustrine plain exhibit an alternating layers structure in which four layers such as medium-coarse sand layer contains cobblestone, medium-fine sand layer, silty-fine sand layer contains sandy loam layer and loam layer exist.

2.2 Landform Features

The basin is rhombus-shaped and the direction of its major axis is northwest–southeast, and its area is estimated to be 13,600 km^2 (including the tableland that was formed in the Tertiary System at the edge of basin) (He et al. 2015; Yuan 2003). The basin is surrounded by mountains, and the terrain of northwestern area is higher than it of southeastern part. In general, the altitude of this district is 1400–1600 m above the sea level, except the northern Tianshan mountain peaks at 4000 m. In the southeast corner of this basin, there is a famous freshwater lake—Bosten Lake. The lake-surface elevation is approximately 1400 m above the sea level (it varies yearly and monthly). At the northwest shore and the southwest side of the lake, there is a large swamp lowland (commonly known as "small lake" which harbors lush reeds). There are large desert areas reaching to the south and east sides of the lake. In the middle of the basin, there is a fine soil flat plain packed with a large number of farmlands.

3 Climate

The Yanqi Basin is located in the inland region of Central Asia and in the transition zone between the Northern Xinjiang (Junggar Basin) and the Southern Xinjiang (Tarim Basin). Its inland position accounts for the continental climate. The entire region is marked by rare precipitation, strong evaporation, dry and windy condition, and great seasonal differences in temperature. The annual average temperature is 7.5 °C. The average temperature is −13.2 °C in January and 23.2 °C in July. The accumulated temperature above 10 °C is 3450 °C. The frost-free period lasts about 170 days during a year (Fig. 2). This condition is similar to that in the northern piedmont of the Tianshan Mountains (Wen 1965). Based on the multi-year data from five meteorological stations (the Baluntai station lies in the mountains, other four stations from Hejing, Heshuo, Yanqi, and Korla lie in the plain area), the annual average precipitation is 50.7–79.9 mm, and about 80–90% of the precipitation occurs during the period from May to September (Fig. 3). Precipitation exhibits regional difference, for example, precipitation in mountain is more abundant than it in the plain area. Annual evaporation is up to 2002.5–2449.7 mm, which is 30–40 times more abundant than annual precipitation. Moreover, the monthly variation of evaporation (Fig. 4) and monthly variation of temperature (Fig. 2) have the same trend. The high

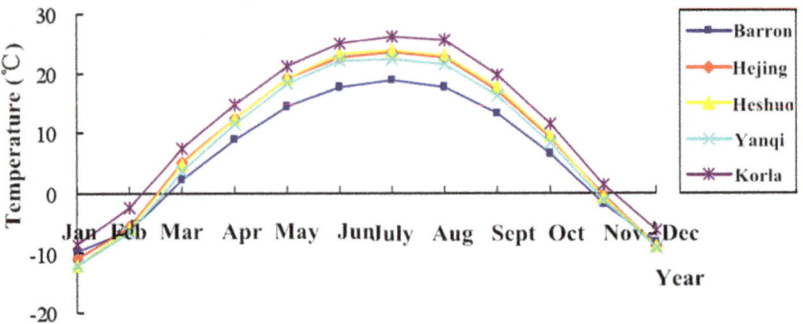

Fig. 2 Monthly average temperature at various stations

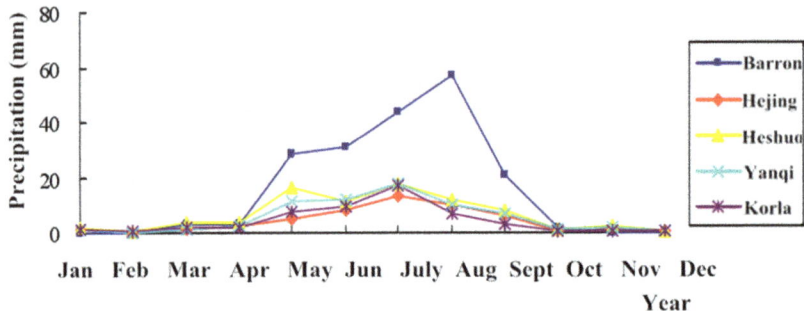

Fig. 3 Monthly average precipitation at various stations

ratio of evaporation: precipitation is a main reason for the deterioration of ecological environment in Yanqi Basin. Snowfall is rare in winter; however, the freezing period lasts for 170 days (from early October to mid-April). The average wind speed is 1.8–2.3 m/s, and the maximum wind speed is up to 5–24 m/s. Windy season occurs in late spring to early summer or in late summer to early autumn, and the northwest wind and southwest wind hold a leading place.

The climate in its surrounding mountainous is much colder and damper than it in the basin. With the increasing of mountain altitude, temperature decreases and the precipitation increases, which exhibits an obvious vertical gradient and owns a lot of natural landscape zones. In the northwestern high/middle altitude mountain zones (with an altitude of >3500 m), mountains are covered by snow year around. The annual precipitation is more than 450 mm in this area. In the western and northern high altitude mountain zones (at an altitude of 2000–3500 m), annual precipitation ranges from 100 to 300 mm. In the southern medium altitude mountains (at an altitude of 2000 m) and low altitude mountains, annual precipitation is only 50–100 mm.

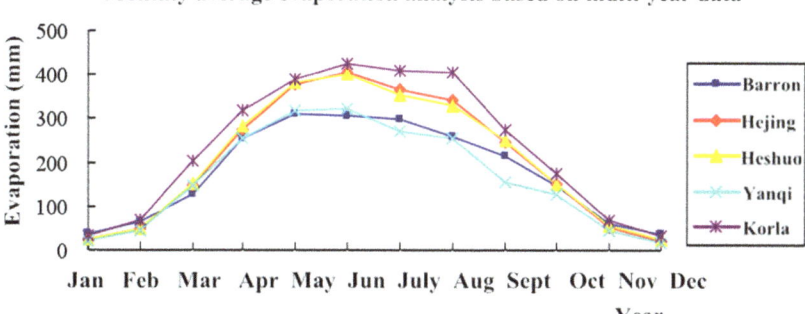

Fig. 4 Monthly average evaporation at various stations

In general, water vapor is transported from northwest to southeast, for which reason climate in northwestern area is much damper than it in southeastern area.

4 Hydrology in the Watershed

Surface runoff in the Yanqi Basin is mainly formed in the mountains, and its distribution is very uneven. The higher the mountain is, the greater precipitation becomes, and the easier surface runoff forms. Thus, in the northwestern high altitude mountain zone and in the northern medium altitude mountain zone, the development of water system is better than it in other zones. In the southern low altitude or medium altitude mountain zones and in southeastern hills, perennial surface runoff is rare. Only after the summer rainstorm occurred, flood current can be formed in valley. Rivers in the plain are originated from the mountains; thus, the distribution of hydrological network in the plain is similar to that in the adjacent mountain areas. For example, the perennial river is only distributed in the western and northern piedmont plains in the basin, and groundwater is recharged by a large amount of seepage water from the Gobi gravel zone, the rest water finally flows into the Bosten Lake; by contrast, there is no river in south and eastern piedmont plain, and only sporadic spring flows are distributed in the adjacent mountainous areas. Spring-fed rivers are formed by the larger spring flows; however, these rivers have vanished in the area not far from the mountain due to strong infiltration.

There are five rivers (the Kaidu River, the Haheren Stream, the Huangshui Stream, the Qingshui River, and the Ustala River), which originate from northwestern perennial snow-clad mountains and flow into the basins (Fig. 5). All of them are mainly recharged by meltwater, precipitation, and groundwater (in the low-precipitation season during winter, rivers are mainly recharged by groundwater). The Kaidu River

Fig. 5 Distribution of the river system in Yanqi Basin

and Huangshui Stream can flow to the central basin throughout the year, and water of these two rivers finally flow into the Bosten Lake (Zhou et al. 2001; Zuo et al. 2004). In other rivers, part of the water is used for irrigation, and most of the water has infiltrated into groundwater in the piedmont Gobi plain and finally flow into the Bosten Lake. Kongque River, which is the only way out of the entire basin, lies on the southwest of Bosten Lake. It flows toward southwest and passes through the Iron Gate Pass. Some water of the lake discharges out of the basin and flows into the Korla area. The annual runoff of three rivers (the Kaidu River, Huangshui Stream, and Qingshui River) accounts for 93% of the total runoff (39.71×10^8 m^3) in Yanqi Basin, of which the annual runoff of the Kaidu River accounts for 84.7%.

The Kaidu River is the largest river flowing through the Yanqi basin. The total length is 560 km, and its watershed area is 4.43×10^4 km^2. Its average annual runoff is 34.82×10^8 m^3, and the average amount of diversion water from the Kaidu River in irrigation area is 10.93×10^8 m^3. On average, the amount of water flowing into lake is 21.44×10^8 m^3 per year. It receives water inflow from a huge catchment area, and the water is mainly recharged by rainfall and meltwater from high mountains. Distribution of water flow in the summer half-year is higher than it in the winter half-year. Water quality is good and belongs to the HCO_3-Ca type, and the water mineralization degree is 0.2–0.3 g/L. Moreover, the river bed is not deep from the surface and is easy to divert water, for which reason it is the main water resource for irrigation in the basin. Flood discharge is very large in July–August (accounting for 32% of the input water in the whole year), causing a rise in the water level of river and Bosten Lake and often drowning farmland in the lake area.

Bosten Lake is the largest freshwater lake in Xinjiang, and it is also the final converging place for surface water and groundwater in Yanqi Basin. Water in the Bosten lake is mainly derived from the Kaidu River (Zhang and Fu-Hua 2004), followed by discharged water and groundwater from the Huangshui Stream and irrigation area. The east–west length of the lake is 50–60 km, and the south–north length is 20–25 km, and the water area is about 1000 km^2. The lake water level peaks during the period from April to May and reaches the lowest value during the period from November to next February. Lake water level varies in the range of 0.4–0.6 m. Recently, due to a large amount of water diverted for irrigation, lake water level has decreased year by year. In addition, a significant amount of lake water is consumed through evaporation (10.86 × 10^8 m^3/a). The water mineralization degree of Bosten Lake is 1.1–1.4 g/L, and the water belongs to the SO_4-CO_3-Na-Mg type and is in slight sulfate mineralized stage. The evapotranspiration in the western reedy marsh is 8.86 × 10^8 m^3/a. The reedy marshland, spanning an area of 662 km^2, lies on the west of the Lake. There are lots of scattered small lakes over the place, and the total area of the adjacent small lakes is approximately 60 km^2. Between the small lakes and the big Lakes, there are some distributary channels, for which reason the water flow can be auto-regulated. However, in recent years, due to the development of agricultural irrigation, a large amount of salt from the soil has been carried into Bosten Lake through farmland drainage. As a result, water mineralization increases in some areas of the lake and has reached 10 g/L (Dong et al. 2006).

5 Groundwater Characteristics

In the Yanqi Basin, the storage of groundwater in the pores of Quaternary ravelly ground is estimated to be more than 1320.5 × 10^8 m^3. This superior natural condition makes the Yanqi Basin to become a unique hydrogeological basin with huge natural underground reservoir. But it is not a closed basin because some water from the Bosten Lake is outputted by the Kongque River. This river flows through the Iron Gate Pass and the Korla area and finally flows into the Taitema Lake and Lop Nor that are lie in lower reaches of the Tarim River.

The surface water and groundwater are abundant in the basin, and the annual groundwater recharge is 12.17 × 10^8 m^3 (Zhong 1989). There are very close relationships between surface water and groundwater, such as they can interconnect and interconvert. Groundwater is recharged by river water seepage in the Gobi area, while groundwater overflows in the low terrain area and then forms surface water. Water in river and spring is used for irrigation by utilizing open channel, while the diversion water and irrigation water are formed into groundwater again due to infiltration. Owing to the amount of groundwater recharge increasing in the irrigation area, the groundwater level raises at the same time. On the one hand, the groundwater is consumed through vertical evaporation and plant transpiration; on the other hand, it is discharged into the runoff in low-lying areas (such as rivers, canals, lakes) and then forms surface water. Under natural conditions and human engineering activities, the contact and mutual conversion between surface water and groundwater can balance the varying trend and range of groundwater level.

Groundwater is greatly consumed in the basin through a surface evaporation (Phreatic water is consumed through soil evaporation and plant transpiration), and the rest part is discharged into the Bosten Lake, reedy marsh, and peacock river. In the canyon, the Kongque River flows through the bedrock, which results in conversion of all underground runoff into surface runoff. Thus, the surface water and groundwater in the basin form an unified water body that keeps circulating and exchanging, which essentially makes the whole water body remain freshwater nature in the basin.

From the piedmont to the central basin and to the Bosten Lake, the chemical characteristics of groundwater present regular change in horizontal zone. In the piedmont zone, alluvial–diluvial fan or diluvial fans are widely distributed, which constitute piedmont inclined cobbly Gobi plain. Phreatic aquifer is composed of a single thick sandy gravel, and the layer thickness is 200–300 m. The buried depth of underground water is 8–50 m. The aquifer has great thickness, abundant water, and fine quality. Degree of water mineralization is generally less than 1 g/L. Most of the water belongs to HCO_3-Ca or HCO_3-Ca-Mg type. Groundwater and river water are mainly recharged by melting snow and rainfall from high mountains and middle altitude mountains. The lower part of the alluvial fan and the lakeside are a fine soil plain. The formation stratum presents an alternating layers structure in which sandy gravel layer, sandy loam layer, and loam layer exist. Phreatic water and confined water are formed in this formation stratum. As the lithologic character in aquifer becomes finer, the water level gradually gets much shallower. The depth of groundwater is generally less than 3 m in the plain area. Under strong evaporation, groundwater continuously upward moves due to the soil capillarity, for which reason the salt in water is brought to the soil surface and leads to soil salinization. Phreatic water is concentrated by evaporation, which results in water mineralization increasing (generally 3–10 g/L). In the area where water table is less than 2 m, the evaporation is much stronger and the soil salinization is more common. In the surrounding area of the Bosten Lake, the marshland and the low-lying depression between fans where the water table is less than 1 m, the water mineralization degree is as high as 10–50 g/L, which result in a mineralization center formed and severe salinization occurred (Huang et al. 1985; Lin and Jin 2006; Liu 1990).

6 Soil Types

The main soil types of the Yanqi Basin include brown desert soil, alluvial soil, and cultivated soil from irrigation—automorphic ancient oasis and irrigation—hydromorphic ancient oasis, meadow soil, bog soil, and solonchak. Solonchak and meadow soil are widely distributed. The distribution of soil in the basin is closely related to the geological, geomorphologic, and hydrogeological conditions. Two most obvious changes can be found from northwest to southeast and from north to south in the direction. From the northwest to the southeast, there are piedmont, the Kaidu river ancient delta (upper, middle, lower, and edge), Kaidu River modern delta and Bosten

Lake. From north to south, there are piedmont, proluvial–alluvial fan (upper, middle, and lower part), fan margin and littoral barrier plain of Bosten Lake.

6.1 Brown Desert Soil

The brown desert soils are distributed in the upper parts of proluvial–alluvial fan, and the parent material mainly consists of cobbly stone. In the 10–50 cm soil layer, soil is mixed with sandy loam and coarse sand. In this area, cobbly brown desert soil is mainly distributed in newly formed diluvial fan and lies in the part from Heshuo to Quhuizhuang, while the cobbly-gypsum brown desert soil is distributed in the rest part.

6.2 Cultivated Soil from Irrigation—Automorphic Ancient Oasis

Cultivated soil from irrigation—automorphic ancient oasis includes two classes: irrigation—automorphic ancient oasis cultivated soil (White clay) and irrigation—automorphic ancient oasis mature cultivated soil (Loess). The clay-type soil is mainly distributed in the area of Wushitala, Quhuizhuang, Halamaodu, Tahaqi, and Hejing. This kind soil is characterized with the following features: lighter texture, pretty dry, hardening surface soil, lacking of organic matter, and the surface appears grayish white and/or yellowish white. This kind soil is used to plant barren-tolerance flax and corn and so on, but the yield is low. In contrast, Loess-type soil has higher organic matter content and better plow layer structure and soil permeability. Loess-type soil is widely distributed and used to grow corn, wheat, and other crops.

6.3 Cultivated Soil from Irrigation—Hydromorphic Ancient Oasis

From the soil profile, that there is about 20-cm thick plow layer on the top, in which the soil appears grayish black and has granular structure and better permeability. The soil below the plow layer presents steel gray mix with rust spots. This type of soil is used to grow corn and wheat. However, salinization is common in this soil, which mainly arises in interfluve lowland lies in middle/lower Kaidu River Delta and in some parts of fan margin.

6.4 Meadow Soil

There are two types of meadow soils in the Yanqi Basin: alluvial meadow soil and fan-land meadow soil. Meadow soil is mainly distributed in the dorsal part of the Kaidu River Delta and in the fan margin lies on the west of Hejing and to the north of the Kaidu River. The groundwater level is generally 1–3 m, and the mineralization degree is less than 1 or 1–3 g/L. Meadow soil is covered with abundant vegetation which is mainly made up of *Achnatherum* spp. accompanying *Glycyrrhiza* spp. and *Alhagi sparsifolia*, and so on. The thickness of humus horizon in meadow soil is about 20 cm, and the content of organic matter is between 1.5 and 2%. Salinization is often seen in most meadow soils, and average salt content of 0–30 cm soil is 0.5–1.0%. In the area adjacent to the modern delta, the formation of the dark meadow soil is the result of the de-swamping. When the Kaidu River is eroded downward and the modern delta extended forward, the groundwater level reduces to 2–3 m, then the runoff condition becomes better and the soil gradually develops into meadow soil. Besides the general characteristics of the meadow soil, dark meadow soil also consists of buried dark gray and charcoal gray peat horizon.

6.5 Bog Soil

Bog soils are distributed in the low-concave parts of the fan margin which lies to the west of Hejing and in the west bank of Bosten Lake. The groundwater level is generally less than 1 m. Bog soils are covered with dense reedy community. The salinization is widely occurred in the Yanqi Basin. Salt content of the 0–30 cm soil layer is 0.5%, and the soda content is significant high [total alkalinity (HCO_3^-) is above 0.17 mmol/L].

6.6 Solonchak

Solonchak occupies a wide area in the basin. Vegetation types and plant growth conditions are closely related to soil salinity. Plant species in meadow solonchaks area is mainly composed of *Achnatherum splendens* accompanying *Glycyrrhiza* spp. and *A. sparsifolia*; plant species in orthic solonchak area is mainly composed of *Tamarix* spp., *Halostachys caspica*, *Haloxylon ammodendron*, *Nitraria* spp., and so on.

Solonchaks are distributed in the edge of the delta and in the fan margin. Abundant groundwater recharge source combined with the backwater effect of Bosten Lake leads to runoff blocked, for which reason the conditions of soil reclamation is poor in this area. The content of solonchak in the basin is 2–20% in the 0–30 cm soil layer. Chloride–sulfate are the main components of salt in most solonchaks. Except for a few chloride orthic solonchak, the soda salinization is common in solonchaks. The total alkalinity of surface soil is more than 0.05–0.5%.

7 Vegetation

The Yanqi basin is mainly composed of shrubs desert, succulent halophytic desert, herbosa (*A. splendens*), and reedy marshes (Zhang et al. 2006). Piedmont alluvial fans consist of gravel Gobi belt and sandy gravel Gobi belt; thus, vegetation is rare in the marginal area. Plant species of desert vegetation are as follows: *H. ammodendron*, *Anabasis* spp., *Corchorus capsularis*, *Calligonum mongolicum*, *A. sparsifolia*, *H. caspica*, *Halocnemum strobilaceum*, *Tamarix* spp., *Lycium* spp., *Tamarix* spp., *Kalidium foliatum*, *Reaumuria songonica*, *Nitraria* spp.; plant species of meadow vegetation are as follows: *Iris lacteal* and *Apocynum venetum*; plant species of swamp vegetation are as follows: *Scirpus validus*, *Typha orientalis*, *Phragmites australis*, and so on (Wu 1980).

The occurrence and distribution of plants in Yanqi basin are closely related to the conditions of water and salt distribution. The desert belts which surrounding marginal irrigated area mainly consist of *Suaeda* spp. community and *Halogeton glomeratus* community; the riverside in the irrigated area is mainly composed of *Tamarix ramosissima* community; the central lakeside zone is mainly made up of *P. australis* community. *Suaeda* spp. is the dominant species and constructive species of plant communities in the desert. The abundance and evenness of desert plants are low in marginal area, and it significantly varies with the changing of water and salt distribution. In the marginal area, the most prominent feature of vegetation is there are xerophytic and a large number of halophilic vegetation exists. Such plant species have the characteristics of salt secretion, water storing, high-osmotic pressure, foliage succulent, branches and leaves extremely downsizing, and so on. The life form of the plant community is dominated by xerophytic shrubs, subshrubs, and succulent plants. They form sparse plant communities and the representing species including *Suaeda* spp. (the dominant species), *Nitraria* spp., and *H. glomeratus*. In riverside, *Tamarix* spp. become constructive species, and more herbaceous plants are distributed in this area (Wang 2012).

8 The Social Population Situation

The land in both sides of the Kaidu River is fertile, and the climatic conditions are fine. Thus, this place is known as a production base of crops, cotton, and sugar beet in Xinjiang. In addition, Bosten Lake is one of the two fishery bases in Xinjiang. There is no doubt that Yanqi Basin is important to the agricultural development in Xinjiang. The population in the watershed area is relatively dense, and the Yanqi

Basin has a population of 499 thousand according to the census in 2011. The basic situations of all counties in the Yanqi Basin are as follows:

The Yanqi Hui Autonomous County lies to the southern Tianshan Mountains and in central Basin. It has a total area of 2570.88 km^2. It currently comprises of 4 townships, 4 towns, 3 state-operated farmlands, and 46 administrative villages. The total population is approximately 160 thousand. Kaidu River passes through the county and has a huge hydropower and water resources. In the upper reaches of the river, there is a hydropower station named Dashankou, which provides adequate power security for the economic development in this region. Yanqi Basin also has considerable petroleum reserves, and "Baoliang Oilfield" is located in the Yanqi County. The measured mineral resources include petroleum, natural gas, coal, and other dozens of kinds with massive reserves; wild medicinal plants include more than 100 species, such as *Glycyrrhiza* spp., *Lithospermum erythrorhizon*, *Codonopsis pilosula,* and so on. Finally, the Yanqi County has a rich and unique tourism resource.

The Hejing County lies to the central-southern part of the Tianshan Mountain and in the northwest of Bayingolin Mongol Autonomous Prefecture. Its total area is 3.49×10^6 hm^2, of which the mountain area is 3.23×10^6 hm^2, accounting for 92.6% of the total area of the county. Plain area is 2.58×10^5 hm^2, accounting for 7.4% of the total area. The total population is 194 thousand in the Hejing County.

The Heshuo County lies in the east of Yanqi Basin and on the south of Tianshan Mountain. It contains an area of 12,892 km^2. According to the current census, it has a population of 75,000. This county is surrounded by mountains on three sides, and one side is adjacent to Bosten Lake. Midwest area is a low-lying plain. Agriculture is mainly dominated by crop cultivation including wheat, corn, rice, sugar beet, cotton, and other crops and by animal husbandry cultivation. Bosten Lake lies on the edge of the county. In particular, Heshuo is located at 42° north latitude. It is suggested that Heshuo has the suitable water and soil conditions for wine grape cultivation, and thus it has broad prospect for wine industry development.

The Bohu County lies in the eastern part of Yanqi Basin and in lower reaches of Kaidu River. It contains an area of 1646 km^2, of which the water area is 1646 km^2 (accounting for 43.2% of the total area). It has resources advantage, including aquatic products, reed, tourism, and petroleum. This area is also known for its moderate and humid climate and for fertile land, and all the conditions are suitable to develop agriculture and animal husbandry. It has a total population of 61,000. Bosten Lake is the largest fishery production base in Xinjiang, and own 32 species of freshwater fish that are artificially stocked and naturally breed. Reedy area around the lake is 4×10^4 hm^2, and it is one of the four major reedy areas. Its annual reedy reserve is more than 2×10^5 t. The existing cultivated land area is 3641.3 hm^2 in the Bohu County.

References

Dong X, Zhong R, Liu F (2006) Differentiation of water-salt-interaction in Bostan lake and Peacock river in recent fifty years. Sci Technol Rev 24:34–37

He WG, Wu CH, Zhao J (2015) Study on natural distribution and biological characteristics of *Lycium ruthenicum* in Yanqi Basin. Mod Agric Sci Technol 13:91–93

Huang Y, Jiang DQ, He YK (1985) The comprehensive control of Yanqi Basin alkaline-saline soil and the protection of ecological environment in Bostan Lake. Sci Agric Sin 18:1–8

Kou W, Wang SX, Qian Y (2008) Emergy analysis of oasis agricultural economic ecosystem in Yanqi Basin. J Xinjiang Agric Univ 15:97–101

Lin LR, Jin MG (2006) Characteristics of hydrological cycle and salt migration in Yanqi Basin. Yellow River 28:5–7

Liu ZY (1990) Movement of water and salt in Yanqi irrigation area and prevention approach of salinization. China Rural Water Hydropower 8:15–19

Wang SX (2012) Water and soil resources in the Bosten Lake basin development and sustainable development. China Water Power Press

Wen ZW (1965) Xinjiang soil geography. Science Press, Beijing

Wu ZY (1980) Chinese vegetation. Science Press, Beijing

Yuan ZW (2003) Analysis of tectonic evolution in Yanqi Basin. J Jianghan Petrol Inst 25:33–35

Zhang LC, Fu-Hua N (2004) Agricultural exploitation and water resources change in Bositeng Lake region. J Arid Land Resour Environ 36–40

Zhang J, Zhou CH, Jian xin LI (2006) Spatial pattern and evolution of oases in the Yanqi Basin, Xinjiang. Geogr Res 25:350–310

Zhong XC (1989) Groundwater development in Yanqi Basin from phreatic evaporation. Xinjiang Agric Sci 4:13–15

Zhou CH, Luo GP, Li C (2001) Environmental change in Bosten Lake and its relation with the oasis reclamation in Yanqi Basin. Geogr Res 20(1):14–23

Zuo QT, Jun xia MA, Chen X (2004) Study on variety trend and control about mineralization degree of Bositeng Lake. Adv Water Sci 15:307–311

Climate Change Over the Past 50 Years in the Yanqi Basin

Fengqing Jiang, Junyi Wang and Xiujun Wang

Abstract There has been evidence of warming and significant change in precipitation in northwest China, implying climate change in the vast arid/semiarid regions. This study showed that there was an increasing trend in air temperature over the period of 1960–2014 in the Yanqi Basin, and the warming was most pronounced since the mid-1990s. The warming rate varied over space and between seasons, with a greater rate in autumn (0.30–0.40 °C/10a) and winter (0.29–0.45 °C/10a) than in spring (0.13–0.26 °C/10a) and summer (0.17–0.24 °C/10a). There was also large interannual to decadal variability in precipitation in the Yanqi Basin and runoff in the Kaidu River. Precipitation showed an overall small increasing trend over the period of 1960–2014. The lower reaches of the Kaidu River experienced a significant increase in runoff since the mid-1990s, which might be primarily a result of warming that enhanced melting of snow and glacier in the surrounding mountains.

1 Introduction

There have been numerous studies of temporal variations of air temperature at various spatial scales (Brown et al. 2008; Hansen et al. 2002; Jones et al. 1999; Solomon 2007; You et al. 2011), which show a general warming trend in the global mean air temperature. Earlier studies indicated that the magnitude of warming in the Northern Hemisphere (0.30 °C/10a) was more than the double of the one (0.13 °C/10a) in the Southern Hemisphere during 1977–2001 (Jones and Moberg 2003; Luterbacher et al. 2004).

F. Jiang (✉)
State Key Laboratory of Desert and Oasis Ecology, Xinjiang Institute of Ecology and Geography, Chinese Academy of Sciences, Urumqi 830011, Xinjiang, China
e-mail: jiangfq@ms.xjb.ac.cn

J. Wang · X. Wang
College of Global Change and Earth System Science, Beijing Normal University, Beijing 100875, China

© Springer Nature Singapore Pte Ltd. 2018
X. Wang et al. (eds.), *Carbon Cycle in the Changing Arid Land of China*,
Springer Earth System Sciences, https://doi.org/10.1007/978-981-10-7022-8_3

There is evidence of difference in the warming trend over space and time in China. For example, the warming rate of air temperature was 0.25 °C/10a for the period of 1951–2004 in China (Ren et al. 2005), but 0.35 °C/10a in northwest China during the period of 1961–2006 (Chen et al. 2010; Klein Tank and Können 2003), which were much greater than the global average. Over the past 50 years, an increase in air temperature with a linear tendency of 0.28 °C/10a was observed in Xinjiang, which was lower than that for northwest China (Li et al. 2011). These findings indicated that climate change in Xinjiang might have its own spatial and temporal characteristics due to its large extent and complex terrain.

Apart from the warming trend, there has been evidence that most regions in the world have experienced an increase in precipitation over the last few decades in the last century. Studies have showed significant changes in precipitation in the majority of China, with an overall increase in northwest China (Wang et al. 2004), with implications for the hydrological cycle in Xinjiang. The objective of this study is to investigate the spatial and temporal variability in air temperature, precipitation, and runoff and to explore the possible mechanisms responsible for these changes in the Yanqi Basin.

2 Data and Method

Daily air temperature dataset from four meteorological stations, i.e., the Bayinbuluke, Baluntai, Yanqi, and Kumishi stations, in the Yanqi Basin (including the Kaidu River Basin) with a rough data period from March 1, 1954, to December 31, 2014, were provided by the National Climatic Centre of China (NCCC), China Meteorological Administration (CMA). The quality of the data has been controlled before its release, and the homogeneity test has also been performed. Furthermore, in this study, the double mass curve method was used to check the data consistency (e.g., Li and Yan 2009). The result showed that all the data series in this study were consistent. In total, the missing data account for 0.01% of the data series. The missing data were filled using conventional statistical methods including: (1) If only one day had missing data, the missing data were replaced by the average value of its two neighboring values; (2) if consecutive two or more days had missing data, the missing data would be processed by simple linear regression between its neighboring stations (distance <100 km) without considering the effect of terrain altitude (Jiang et al. 2013). For temporal consistence and facilitating comparison, the period of temperature time series was restricted from January 1, 1961, to December 31, 2014.

In addition, two hydrological stations which are located in the Yanqi Basin and Kaidu River Basin were also chosen for this study (Fig. 1). Among them, the Bayinbuluke station is situated in the upper reaches of Kaidu River and the Dashankou station is in the lower reaches of the river. Monthly streamflow data from 1956 to 2010 of the two stations were obtained from the Xinjiang Administration of Hydrological and Water Resource. Basic information of these stations is given in Table 1, and the locations of the stations are shown in Fig. 1.

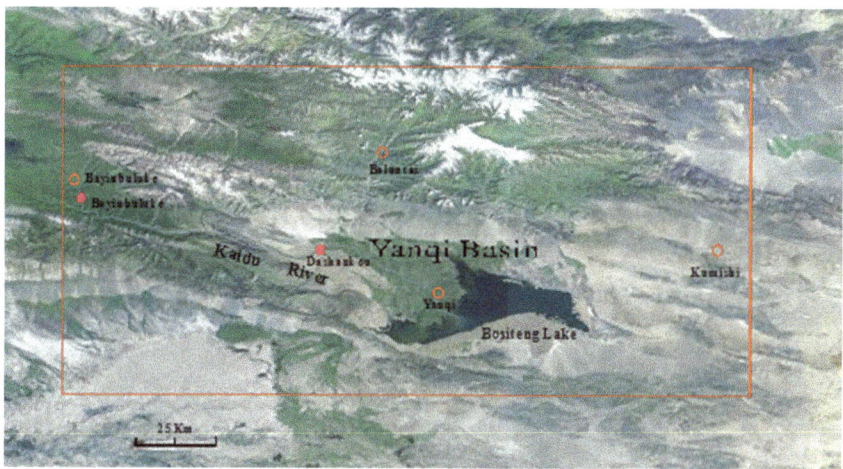

Fig. 1 Sketch map showing the locations of selected meteorological and hydrological stations

Table 1 Basic information of stations chosen for this study

Station	Type	Longitude	Latitude	Altitude (m)	Data period
Bayinbuluke	Weather	84.15	43.03	2458	1961–2014
Baluntai	Weather	86.30	42.73	1739	1961–2014
Kumishi	Weather	88.22	42.23	721	1961–2014
Yanqi	Weather	86.57	42.08	1055	1961–2014
Bayinbuluke	Hydrology	84.12	42.96	2495	1956–2010
Dashankou	Hydrology	85.72	42.25	1400	1956–2010

Ordinarily linear regression was employed to estimate long-term variations and/or trends in the annual and seasonal temperature, precipitation, and runoff. The coefficient of variation (Cv value) is used to reflect the interannual variability, which is calculated as follows: Cv = standard deviation/average.

3 Results and Discussion

3.1 Temporal Changes in Air Temperature

3.1.1 Seasonal Variations in Air Temperature

Viewing from the distribution of monthly average temperature during a year, the temperature in the Yanqi Basin and its neighboring mountains shows a gradual increase from January to July and a slightly faster decrease from August to December (Fig. 2).

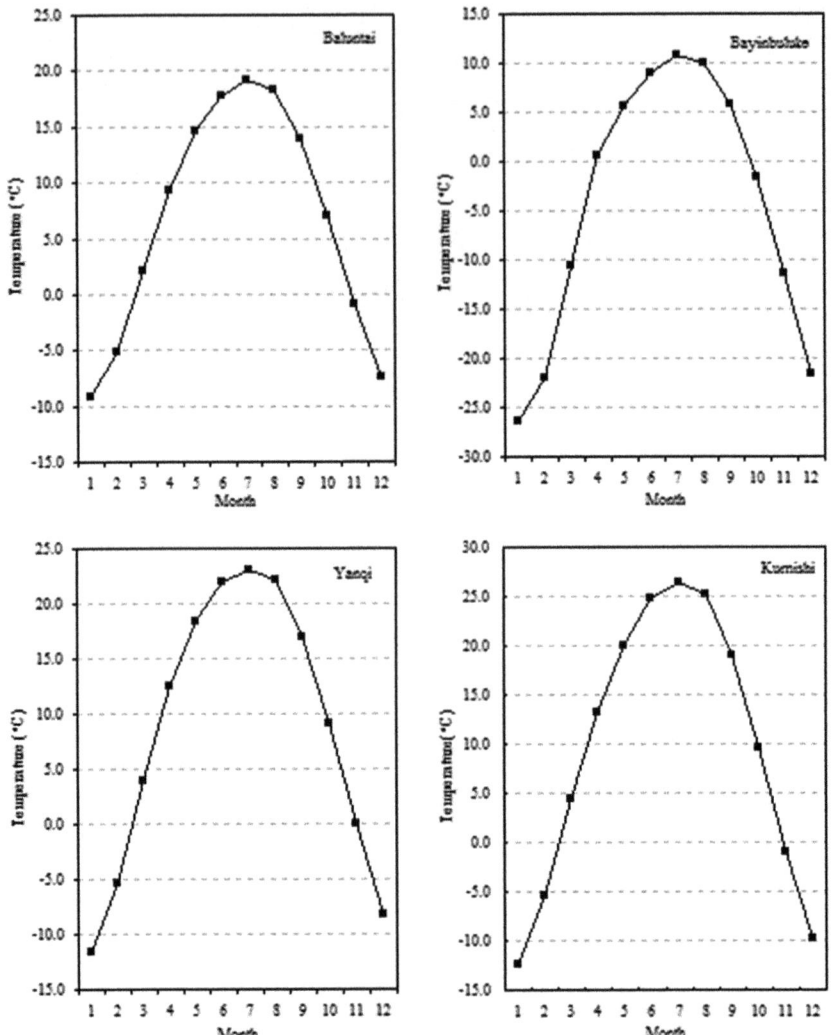

Fig. 2 Climatology of monthly mean temperature at the four weather stations

Overall, the Bayinbuluke station has the lowest temperature year-round, with the mean temperature being −20 °C in winter and 10 °C for summer (Table 2); the Kumishi station reveals highest temperature in all the seasons, i.e., 12.6, 25.5, 9.3, and −6.0 °C for spring, summer, autumn, and winter, respectively.

In general, summer is usually the period with the smallest variability in temperature, while the fall and winter seasons are the periods of largest variability in the Yanqi Basin. However, spring is the period with the highest variability for

Table 2 Mean, standard deviation (Std), and coefficient of variation (Cv) of the temperature for different seasons

Station	Spring			Summer			Autumn			Winter		
	Mean	Std	Cv	Mean	Std	Cv	Mean	Std	Cv	Mean	Std	Cv
Bayinbuluke	−1.5	1.5	0.99	10.0	0.6	0.06	−2.4	1.5	0.62	−20.0	2.6	0.13
Baluntai	8.6	1.0	0.11	18.4	0.8	0.04	6.7	1.2	0.18	−5.1	1.4	0.28
Yanqi	11.6	0.9	0.08	22.4	0.5	0.02	8.7	0.9	0.10	−5.6	1.3	0.24
Kumishi	12.6	0.9	0.07	25.5	0.6	0.02	9.3	1.5	0.24	−6.3	1.5	0.24

Table 3 Decadal changes in mean annual temperature (°C) at the four stations

Station	1960s	1970s	1980s	1990s	2000s	Mean (1961–2014)
Bayinbuluke	−4.5	−4.2	−5.1	−4.2	−3.4	−4.4
Baluntai	6.3	6.1	6.1	6.8	7.8	6.6
Yanqi	8.2	8.1	8.4	8.9	9.2	8.9
Kumishi	9.4	9.0	9.4	9.8	10.1	9.6

the Bayinbuluke station (Table 2). This may be related to the topography of the mountain basin and instability of the local atmospheric stratification.

3.1.2 Decadal Changes in Air Temperature

Our analyses show that the Yanqi Basin experienced a small decrease in temperature from the 1960s to the 1970s and an overall increasing trend in temperature from the 1970s to the 2010s except at the Bayinbuluke station that showed a significant decrease from −4.2 °C in the 1970s to −5.1 °C in the 1980s, followed by a significant increase to −3.4 °C in the 2000s. Among the four stations, the Baluntai station is characterized by its largest increase in temperature, 1.5–1.7 °C from 1961s to 2010s, while the Kumishi station appears a slightly increasing trend in temperature, i.e., 0.5–0.9 °C from 1961s to 2010s (Table 3).

3.1.3 Interannual Variability and Change Trend in Different Seasons

Figure 3 shows large interannual variability in spring mean temperatures at all four stations although there was an overall increasing trend from 1961 to 2014. It appears that there was no clear trend prior to the 1980s, but an increasing trend starting in 1987, which is coincident with the conclusion that a change point of temperature time series could be occurred in the middle of the 1980s (Li and Jiang 2007). The linear tendency rate of spring temperature over the period of 1961–2013 was 0.13, 0.19, 0.26, and 0.16 °C/10a at the Bayinbuluke, Baluntai, Yanqi, and Kumishi stations, respectively, which was not statistically significant (Table 4).

It is shown in Fig. 4 that summer average temperature was increasing from 1961 to 2014 at the four stations. The linear tendency rate of summer temperature was 0.24, 0.17, 0.18, and 0.17 °C/10a at the Bayinbuluke, Baluntai, Yanqi, and Kumishi stations, respectively (Table 4). The increasing trend was significant at the Bayinbuluke station ($P < 0.01$) and the Yanqi station ($P < 0.05$).

Figure 5 shows considerable interannual variation in autumn temperature at all four stations in the Yanqi Basin. In general, autumn average temperature was all increasing from 1961 to 2014. The linear tendency rate in autumn was 0.32, 0.40, 0.33, and 0.30 °C/10a at the Bayinbuluke, Baluntai, Yanqi, and Kumishi stations,

Table 4 Change trend (°C/10a) of mean temperature and its coefficient (R) for different seasons

Station	Spring Trend	Spring R	Summer Trend	Summer R	Autumn Trend	Autumn R	Winter Trend	Winter R	Annual Trend	Annual R
Bayinbuluke	0.13	0.02	0.24	0.46**	0.32	0.11	0.29	0.03	0.17	0.07
Baluntai	0.19	0.09	0.17	0.12	0.40	0.29*	0.45	0.25	0.27	0.26*
Yanqi	0.26	0.20	0.18	0.29*	0.33	0.34**	0.39	0.22	0.25	0.35**
Kumishi	0.16	0.07	0.17	0.20	0.30	0.22	0.26	0.07	0.17	0.18

*$P < 0.05$, **$P < 0.01$

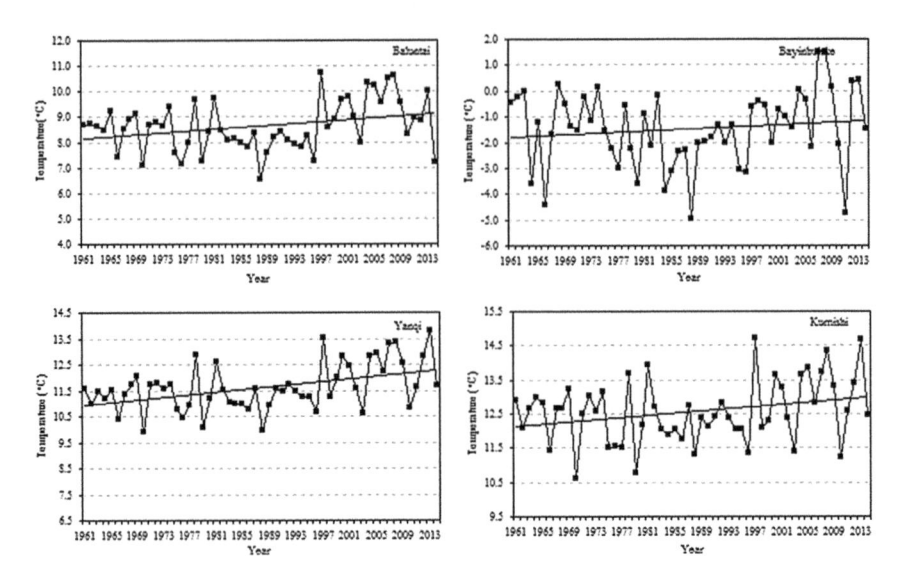

Fig. 3 Interannual variability of temperature and trend in spring at the four stations

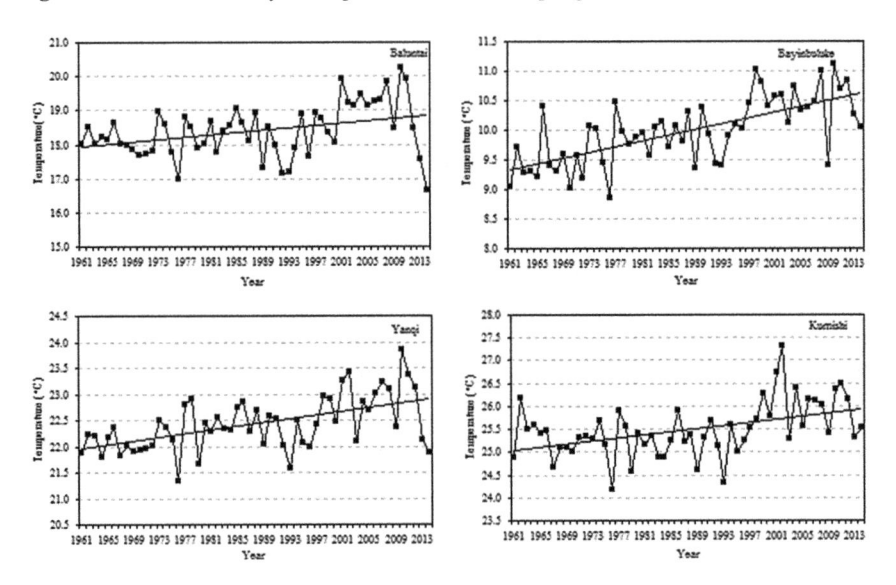

Fig. 4 Interannual variability and trend in summer temperature at the four stations

respectively, which was much greater than those in spring and summer (Table 4). The increasing trend was significant at the Baluntai and Yanqi stations.

There was interannual to decadal variability in winter temperature at all four stations (Fig. 6). Overall, winter average temperatures showed an increasing trend during the period of 1961–2014. The linear tendency rate was 0.29, 0.45, 0.39,

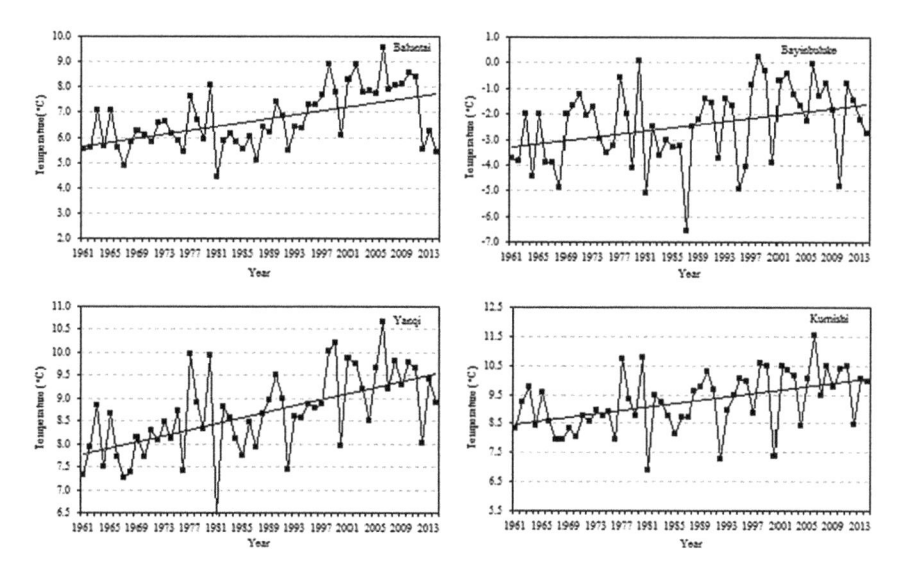

Fig. 5 Interannual variability and trend in autumn temperature at the four stations

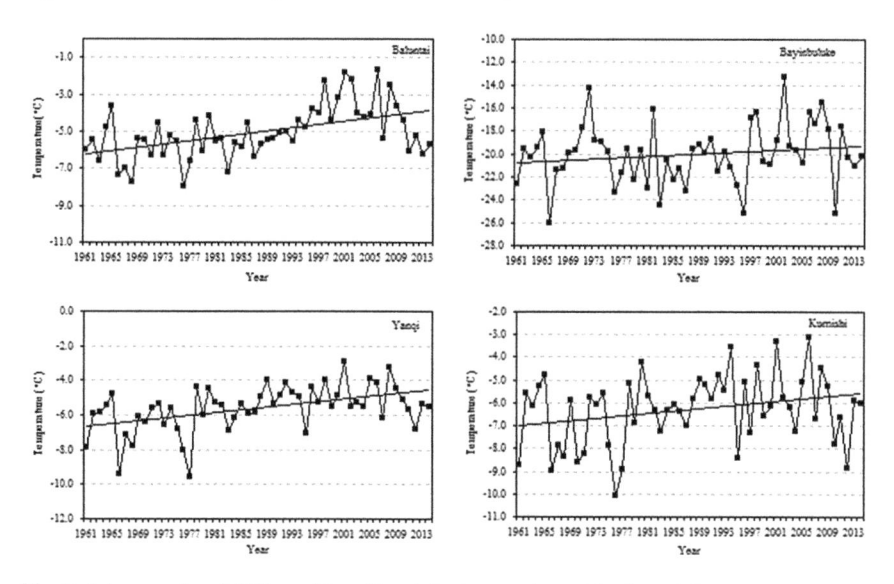

Fig. 6 Interannual variability and trend in winter temperature at the four stations

and 0.26 °C/10a at the Bayinbuluke, Baluntai, Yanqi, and Kumishi stations, respectively (Table 4), although the change trend was not statistically significant. The most obvious warming occurred in autumn and winter, which is in agreement with some previous studies (Long et al. 2016; Wang et al. 2012).

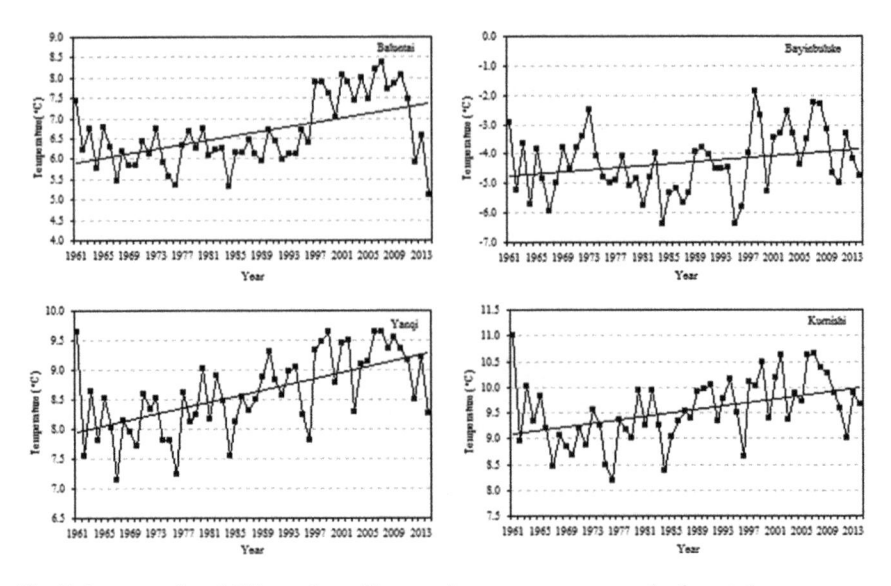

Fig. 7 Interannual variability and trend in annual mean temperature at the four stations

3.1.4 Variability and Trend in Annual Mean Temperature

Figure 7 shows large interannual variations in annual mean temperature at the four stations in the Yanqi Basin. It can be seen that a decreasing trend occurred from the beginning to the late of 1960s, then a gradually increasing trend appeared in the period of 1970–2009 (except at the Bayinbuluke station), and finally, a decreasing trend took place since 2010. Overall, there was an increasing trend for four stations from 1961 to 2014, and the linear tendency rate was 0.17, 0.27, 0.25, and 0.17 °C/10a at the Bayinbuluke, Baluntai, Yanqi, and Kumishi stations, respectively. The increasing trend was significant at the Baluntai ($P < 0.05$) and Yanqi stations ($P < 0.01$) (Table 4).

3.2 Temporal Changes in Precipitation

3.2.1 Monthly Variations in Precipitation

There is strong seasonality in precipitation in the Yanqi Basin, with the largest in July and extremely low rate (<3 mm/month) during the period of November–March at all the stations (Fig. 8). Clearly, precipitation is much higher at the Bayinbuluke and Baluntai stations than at the Yanqi and Kumishi stations. In particular, July's rainfall is greater than 55 mm in the former, but less than 20 mm in the latter.

The variation coefficient of seasonal precipitation is commonly used to characterize intra-annual variability of precipitation in a given region. A smaller variation

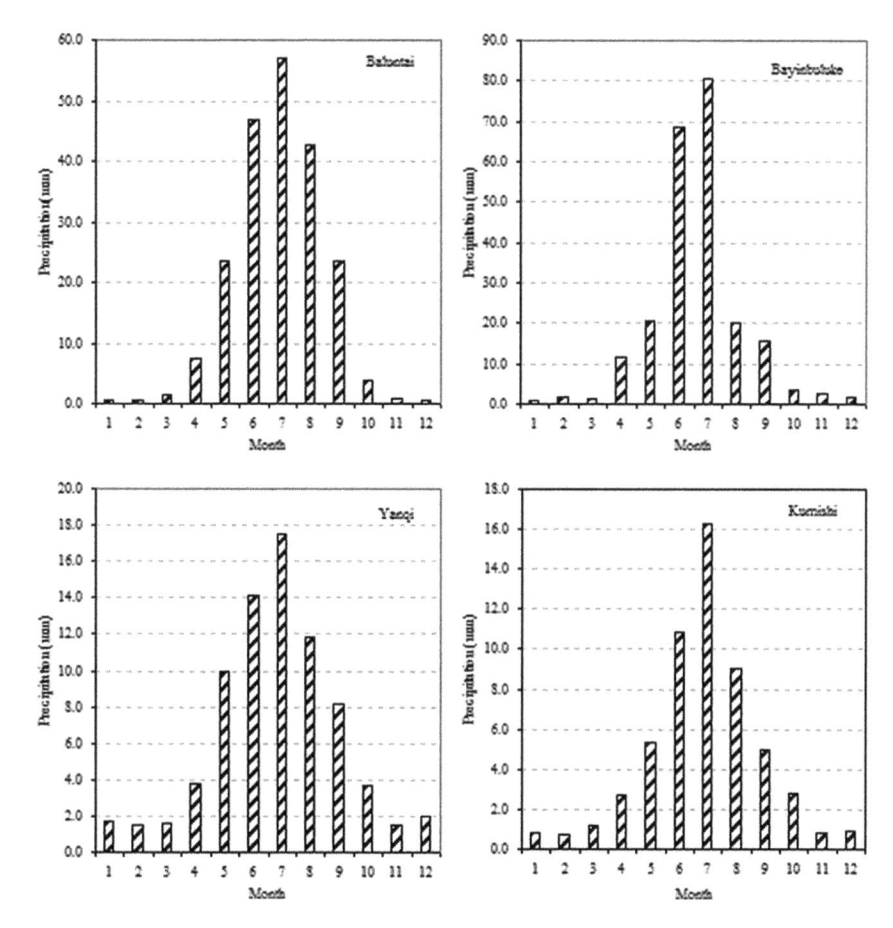

Fig. 8 Climatology of monthly precipitation in the Yanqi Basin

coefficient indicates a smaller interannual variation of precipitation and a higher utilization value of precipitation resources. It is given in Table 5 that in the Yanqi Basin, seasonal precipitation variability was high. Specially, summer is usually the period with the minimum precipitation variability, while the fall and winter seasons are the periods with maximum precipitation variability, demonstrating a greatly interannual fluctuation in precipitation during winter and autumn seasons.

Table 5 Mean, standard deviation (Std), and coefficient of variation (Cv) of precipitation at the four stations

Station	Spring			Summer			Autumn			Winter		
	Mean	Std	Cv	Mean	Std	Cv	Mean	Std	Cv	Mean	Std	Cv
Bayinbuluke	43.3	17.5	0.40	183.8	43.5	0.24	36.8	17.1	0.46	11.8	7.2	0.61
Baluntai	32.7	18.9	0.58	149.1	53.5	0.36	27.9	20.7	0.74	1.7	1.9	1.13
Yanqi	15.2	14.9	0.98	43.9	23.5	0.54	13.0	14.9	1.14	4.9	5.5	1.11
Kumishi	9.2	7.5	0.82	36.5	15.4	0.42	8.2	8.6	1.05	2.2	2.8	1.25

Table 6 Interdecadal changes in mean annual precipitation at the four stations in/near the Yanqi Basin

Station	1960s	1970s	1980s	1990s	2000s	Mean (1961–2014)
Bayinbuluke	261.0	258.7	259.2	281.9	302.0	301.9
Baluntai	190.9	204.5	184.8	253.3	226.2	207.2
Yanqi	72.5	62.3	82.2	96.4	81.1	58.1
Kumishi	47.7	45.6	50.3	74.4	59.6	59.4

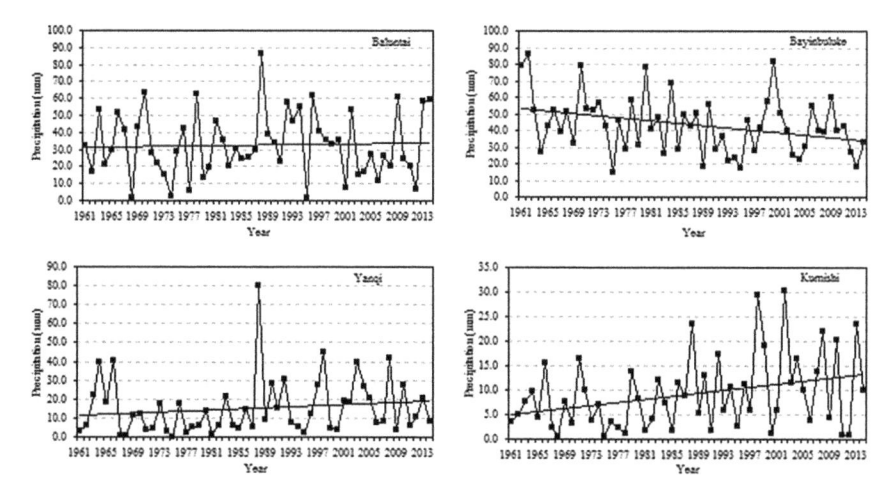

Fig. 9 Temporal variation and trend in spring mean precipitation at the four stations

3.2.2 Decadal Changes in Precipitation

The Yanqi Basin experienced considerably decadal changes in precipitation over the period of 1961–2014. There was a clear increasing trend (from ~260 mm in the 1960s to ~300 mm in the 2000s) at the Bayinbuluke station and an overall increasing trend prior to 2000, followed by a decrease from the 1990s to the 2000s at the other three stations (Table 6).

3.2.3 Trend in Different Seasons

It is shown in Fig. 9 and Table 7 that the change trend in spring precipitation is different among the four stations, i.e., an increasing trend at the Kumishi (1.56 mm/10a), Yanqi (1.40 mm/10a), and Baluntai (0.43 mm/10a) stations, but a decreasing trend at the Bayinbuluke (−3.55 mm/10a) station. The largest variability in spring precipitation is found at the Kumishi station and smallest at the Baluntai station (Table 7).

Table 7 Change trend of mean precipitation and its coefficient (R) for different seasons

Station	Spring		Summer		Autumn		Winter	
	Trend	R	Trend	R	Trend	R	Trend	R
Bayinbuluke	−3.55	0.10	9.62	0.12	2.31	0.05	1.30	0.08
Baluntai	0.43	0.00	9.32	0.07	−0.57	0.00	0.24	0.04
Yanqi	1.40	0.02	−0.45	0.00	−0.15	0.00	0.57	0.03
Kumishi	1.56	0.11	2.06	0.04	0.20	0.00	0.48	0.07

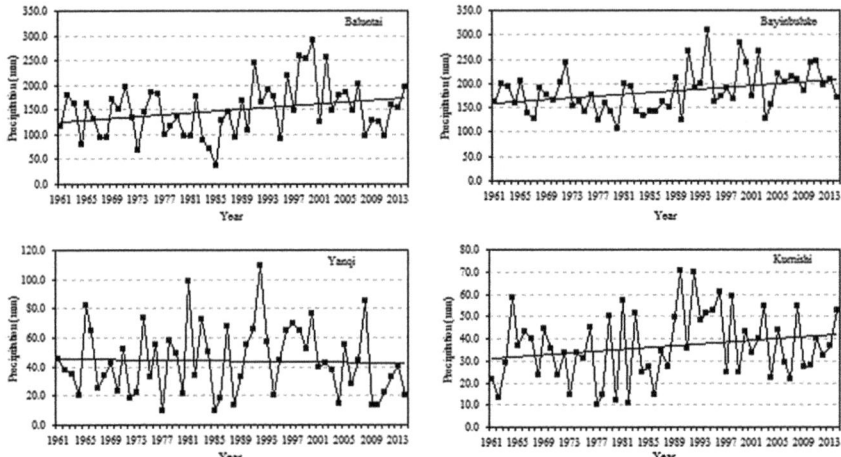

Fig. 10 Temporal variation and trend in summer mean precipitation at the four stations

Figure 10 shows the temporal variation of summer precipitation at the four stations in the Yanqi Basin. It can be seen that there is an increasing trend in summer precipitations at the Baluntai, Bayinbuluke, and Kumishi stations, but a slightly decreasing trend at the Yanqi station from 1961 to 2014. The linear tendency rates of summer precipitation are 9.62, 9.32, −0.45, and 2.06 mm/10a at the Bayinbuluke, Baluntai, Yanqi, and Kumishi stations, respectively (Table 7).

The autumn precipitation shows an increasing trend at the Bayinbuluke and Kumishi stations, but a slightly decreasing trend at the Yanqi and Baluntai stations from 1961 to 2014 (Fig. 11). The linear tendency rates of autumn precipitation are 2.31, −0.57, −0.15, and 0.20 mm/10a at the Bayinbuluke, Baluntai, Yanqi, and Kumishi stations, respectively (Table 7).

As shown in Fig. 12, there was a large interannual to decadal variability in winter precipitation despite an increasing trend at all four stations from 1961 to 2014. The linear tendency rate was 1.30, 0.24, 0.57, and 0.48 mm/10a at the Bayinbuluke, Baluntai, Yanqi, and Kumishi stations, respectively (Table 7).

Figure 13 shows temporal variation in annual mean precipitations at the four stations in the Yanqi Basin. There was a large interannual variability but also an overall increasing trend in annual mean precipitation, indicating a potential climate

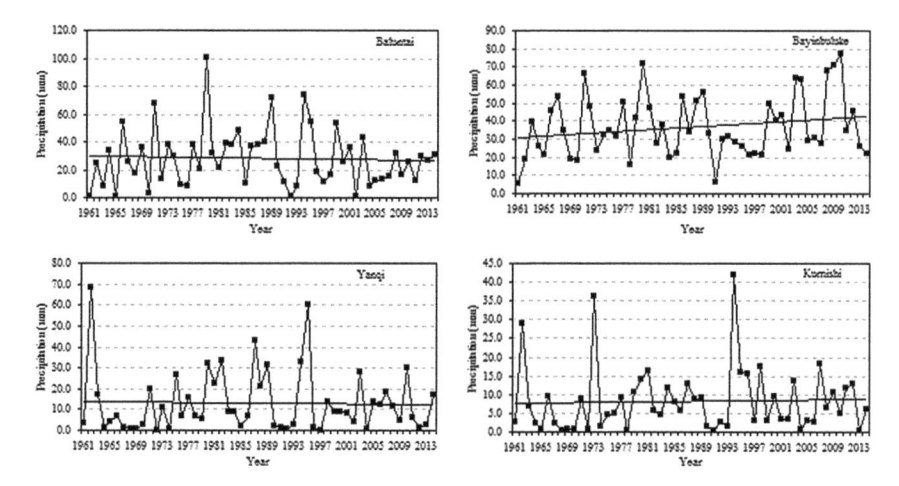

Fig. 11 Temporal variation and trend in autumn mean precipitation at the four stations

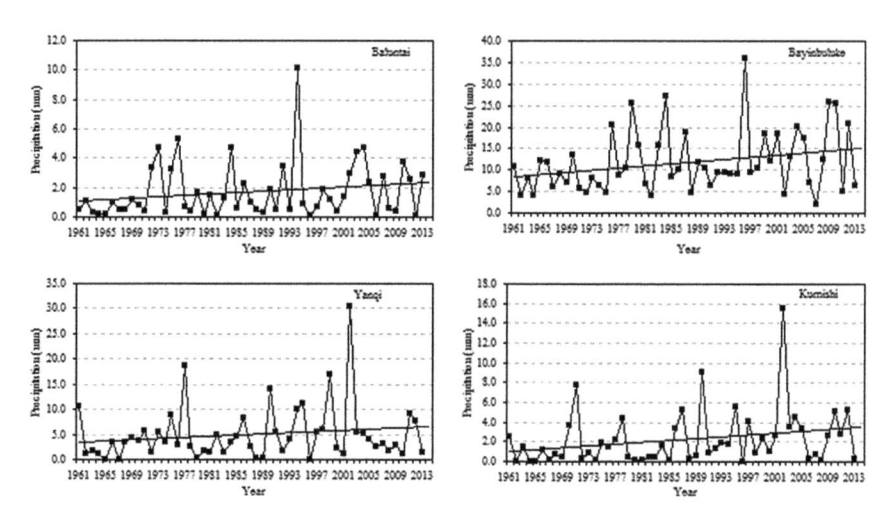

Fig. 12 Temporal variation and trend in winter mean precipitation at the four stations

regime shift from drier climate to wetter climate, which was in agreement with previous studies (Jiang et al. 2013; Shi 2003; Zhang et al. 2012; Zhen and Jiang 2007). The linear tendency rate of annual mean precipitation was 9.73, 8.88, 1.55, and 4.23 mm/10a at the Bayinbuluke, Baluntai, Yanqi, and Kumishi stations, respectively.

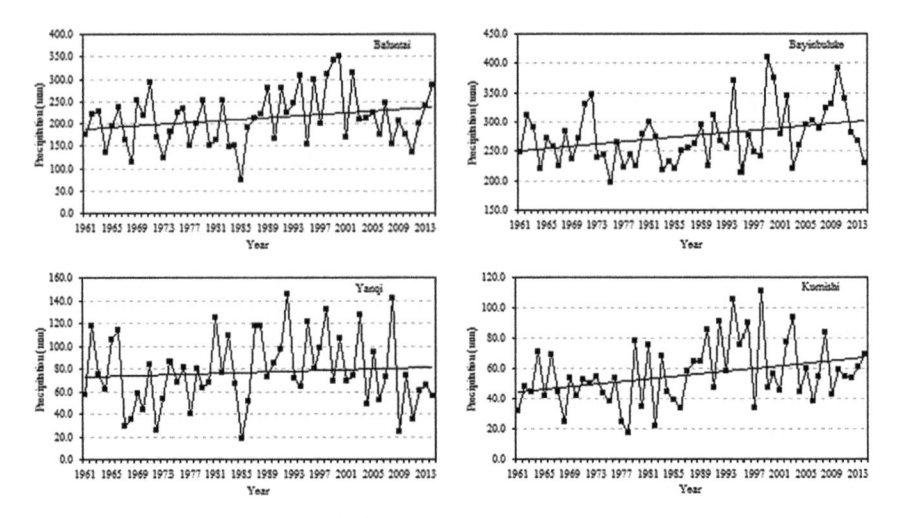

Fig. 13 Temporal variation and trend in annual mean precipitation at the four stations

4 Runoff in the Kaidu River

4.1 Kaidu River: An Introduction

The Kaidu River is an important source of water for the Tarim Basin. The sources of the Kaidu River are located on the central southern slopes of the Tian Shan from where it flows through the Yulduz Basin and the Yanqi Basin into Lake Bosten for which it is the most important tributary (Fig. 1). The full length and the catchment area of Kaidu River is 560 km and 2.2×10^4 km^2, respectively. The Kaidu River Basin belongs to the arid and semiarid climate, and water sources are from natural precipitation, snow, and glacier melting water. Annual mean runoff of the Kaidu River is 34.12×10^8 m^3. Bayinbuluke is the upstream station of Kaidu River, and Dashankou is the downstream station, which controls the total water amount of Kaidu River.

4.2 Statistics of Runoff in Kaidu River

The mean, maximum, and minimum of annual runoff were 23.02×10^8, 38.08×10^8, and 16.03×10^8 m^3 in the upper reaches of Kaidu River for the period of 1956–2010, respectively, and 35.96×10^8, 61.65×10^8, and 24.56×10^8 m^3, respectively, in the lower reaches (Table 8).

The variation coefficient (Cv value) of annual runoff ranges from 0.19 to 0.30, demonstrating a lower interannual variability in the annual runoff of the Kaidu River.

Table 8 Statistics of annual runoff in the Kaidu River

Station	Mean ($\times 10^8$ m^3)	Maximum ($\times 10^8$ m^3)	Minimum ($\times 10^8$ m^3)	Std[a]	Cv[b]
Bayinbuluke	23.02	38.08	16.03	4.53	0.197
Dashankou	35.96	61.65	24.56	8.00	0.222

[a]Std—Standard deviation; [b]Cv—Coefficient of variation

The upstream station (Bayinbuluke) has a smaller Cv value than the downstream station (Dashankou) (Table 8). The main reason for the small interannual variation in the annual runoff of the Kaidu River is the small interannual variation of precipitation and the regulation of glaciers and mountain region (Eziz et al. 2014; Xie and Zhu 2011).

4.3 Variations in Annual Runoff

4.3.1 Variations in Runoff in the Upper Reaches of Kaidu River

In general, the upper reaches of Kaidu River experienced a slightly increasing trend in annual runoff from the 1960s to the 1970s, followed by a decreasing trend from the 1970s to the 1990s and then a rapidly increasing trend from the 1990s to the 2000s. Roughly, the 2000s was the decade with the maximum annual runoff, while the 1980s was the decade with the minimum annual runoff (Table 9).

Figure 14 shows the temporal variation in mean annual runoff in the upper reaches of the Kaidu River. There was a fluctuating upward trend in mean annual runoff during the period of 1956–1971, followed by a rapid downward trend from 1971 to 1977 and then a gradual upward trend during the period of 1977–2010. Generally, there was a decreasing trend in mean annual runoff from 1956 to 2010 at the Bayinbuluke station, and the linear tendency rate was -0.153×10^8 m^3/10a.

Figure 15 reveals the temporal variation of runoff in different seasons at the Bayinbuluke station, the upper reaches of Kaidu River. There was a slightly increasing

Table 9 Decadal variation of annual runoff in Kaidu River (10^8 m^3)

Station	1960s	1970s	1980s	1990s	2000s	Average (1956–2010)
Bayinbuluke (upper reaches)	23.21	24.66	19.57	21.71	25.09	23.02
Dashankou (lower reaches)	32.95	33.12	30.53	36.72	42.77	35.96

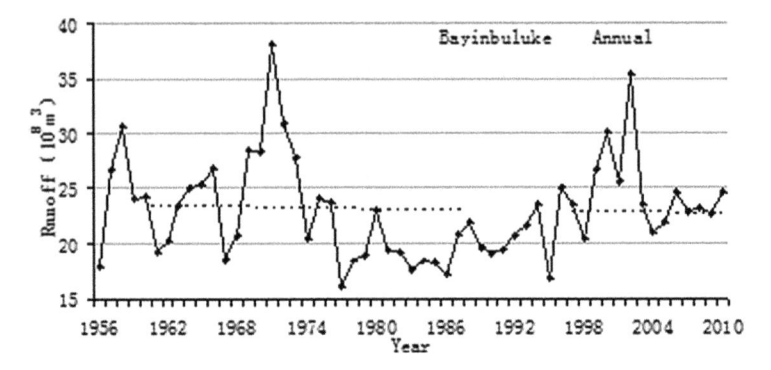

Fig. 14 Variation and trend of annual runoff at the Bayinbuluke station

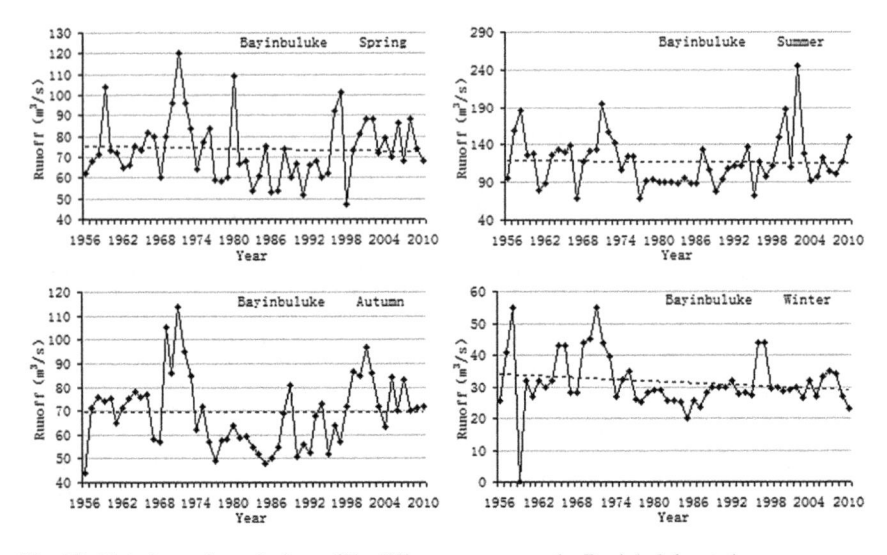

Fig. 15 Variation and trend of runoff in different seasons at the Bayinbuluke station

trend in spring runoff from 1956 to 1971, followed by a decreasing trend from 1971 to 1994 and a rapidly increasing trend from 1994 to 2010. The summer runoff experienced a large fluctuation during the period of 1956–1971, followed by a small fluctuation during the period of 1971–1997 and a fluctuating upward during the period of 1997–2010. The autumn runoff showed a fluctuating upward trend during the period of 1956–1972, followed by a rapid downward trend from 1972 to 1985 and a fluctuating upward trend during the period of 1986–2010. There was a large interannual variability in winter runoff prior to 1960, followed by a clear upward trend during the period of 1960–1972, then a rapid decrease from 1973 to 1985, and a fluctuating upward trend during the period of 1986–2010.

Overall, there were similar downward trends in spring, summer, and winter runoffs. However, autumn runoff showed a slightly increasing trend. The linear

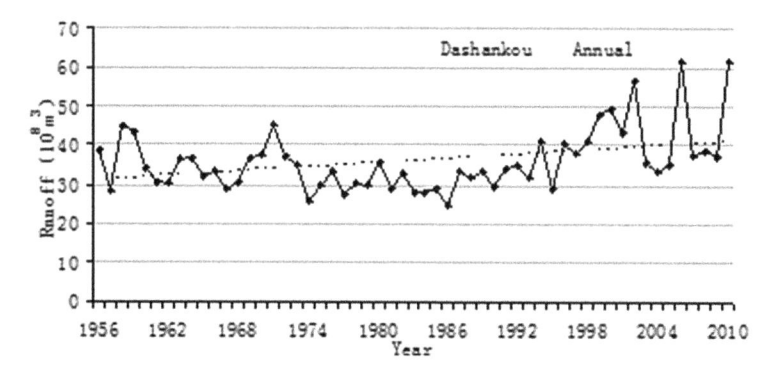

Fig. 16 Variation and trend of annual runoff at the Dashankou station

tendencies of spring, summer, autumn, and winter runoff were -0.15×10^8, -0.48×10^8, 0.002×10^8, and -0.96×10^8 m^3/10a, respectively (Table 10).

4.3.2 Variation in Runoff in the Lower Reaches of the Kaidu River

The lower reaches of the Kaidu River experienced a slight increase in annual runoff from the 1960s to the 1970s, then a decrease from the 1970s to the 1980s, and a rapid increase from the 1980s to the 2000s. The maximum annual runoff was found in the 2000s, and the minimum annual runoff in the 1980s (Table 9).

Figure 16 shows that a fluctuating downward in annual mean runoff in the lower reaches of Kaidu River (Dashankou station) occurred during the period of 1956–1986, and a rapid upward appeared in the period of 1986–2010. Generally, there was an increasing trend in annual runoff from 1956 to 2010 at the Dashankou station. The linear tendency rate of annual runoff was 1.89×10^8 m^3/10a at the Dashankou station. The linear tendencies of spring, summer, autumn, and winter runoff in the lower reaches of Kaidu River were 1.89×10^8, 2.75×10^8, 5.48×10^8, and 6.68×10^8 m^3/10a, respectively (Table 10).

There was a large temporal variation in all four seasons at the Dashankou station (Fig. 17). However, the fluctuation of runoff was more pronounced in spring and summer than in autumn and winter. Overall, the lower reaches of Kaidu River experienced an increase in runoff post-mid-1990s despite the fluctuation. The summer runoff showed a fluctuating downward trend during the period of 1956–1985, followed by a fluctuating upward trend during the period of 1986–2002 and then much lower values with a slightly increasing trend during the period of 2002–2010. The autumn runoff experienced a small change during the period of 1956–1985, but a gradual upward trend from 1986 to 2000, followed by a remarkable fluctuation (between ~80 and ~280 m^3/s) over the period of 2001–2010. The winter runoff showed a similar temporal variation with less degrees of fluctuation to autumn runoff,

Table 10 Trend (10^8 m^3/10a) of annual runoff and its coefficient (R) for different seasons

Station	Annual Trend	Annual R	Spring Trend	Spring R	Summer Trend	Summer R	Autumn Trend	Autumn R	Winter Trend	Winter R
Bayinbuluke (upper reaches)	−0.15	0.00	−0.48	0.00	−0.52	0.00	0.00	0.00	−0.96	0.03
Dashankou (lower reaches)	1.89	0.14	2.75	0.05	5.48	0.03	8.95	0.14	6.68	0.26

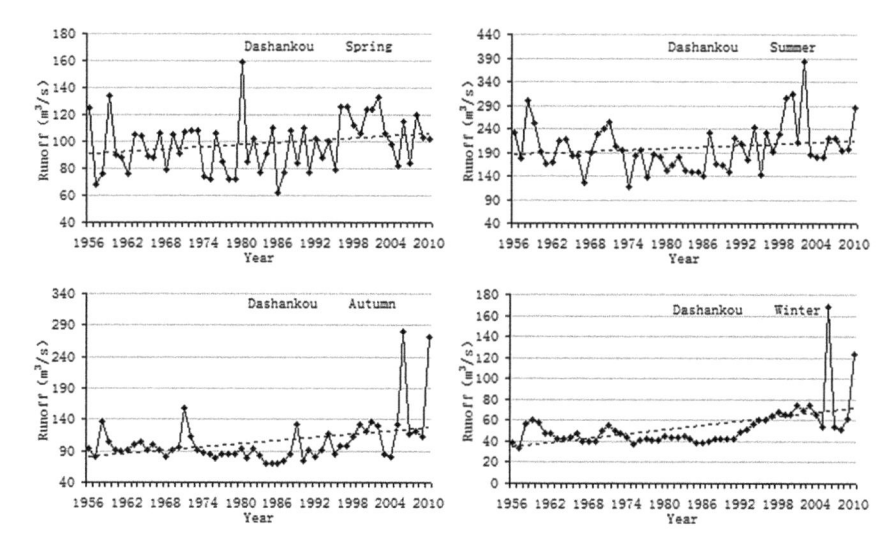

Fig. 17 Variation and trend of runoff in different seasons at the Dashankou station

with a little change during the period of 1956–1990 and an upward trend during the period of 1990–2000, and a large fluctuation during the period of 2000–2010.

5 Conclusions

This study demonstrated an increasing trend (0.17–0.27 °C/10a) in air temperature during the period of 1960–2014 in the Yanqi Basin. However, there were some differences in the warming rate between seasons and over space. In general, warming was more pronounced in the areas of low elevation relative to high elevation; warming rate was much greater in autumn (0.30–0.40 °C/10a) and winter (0.29–0.45 °C/10a) than in spring (0.13–0.26 °C/10a) and summer (0.17–0.24 °C/10a). The greatest warming was observed since the mid-1990s.

There was large interannual to decadal variability in precipitation in the Yanqi Basin, which showed an overall small increasing trend over the period of 1960–2014. The lower reaches of the Kaidu River experienced a significant increase in runoff since the mid-1990s, which might be linked with the warming. Apparently, increasing temperature would lead to enhanced melting of snow and glacier in the surrounding mountains, resulting in more runoff in the Kaidu River.

References

Brown SJ, Caesar J, Ferro CAT (2008) Global changes in extreme daily temperature since 1950. J Geophys Res 113:115–125. https://doi.org/10.1029/2006jd008091

Chen SY, Shi YY, Guo YZ, Zheng YX (2010) Temporal and spatial variation of annual mean air temperature in arid and semiarid region in northwest China over a recent 46 year period. J Arid Land 2:87–97

Eziz M, Yimit H, Rong MA (2014) Analyzing the variation and driving forces of the runoff in Yanqi Basin, Xinjiang during 1956–2010. J Glaciol Geocryol 36:670–677

Hansen J, Ruedy R, Sato M, Lo K (2002) Global warming continues. Science 295:275. https://doi.org/10.1126/science.295.5553.275c

Jiang FQ, Hu RJ, Wang SP, Zhang YW, Tong L (2013) Trends of precipitation extremes during 1960–2008 in Xinjiang, the Northwest China. Theoret Appl Climatol 111:133–148

Jones PD, Moberg A (2003) Hemispheric and large-scale surface air temperature variations: an extensive revision and an update to 2001. J Clim 16:206–223

Jones PD, New M, Parker DE, Martin S, Rigor IG (1999) Surface air temperature and its changes over the past 150 years. Rev Geophys 37:173–199. https://doi.org/10.1029/1999rg900002

Klein Tank A, Können G (2003) Trends in indices of daily temperature and precipitation extremes in Europe, 1946–99. J Clim 16:3665–3680

Li Z, Jiang FQ (2007) A study of abrupt climate change in Xinjiang region during 1961–2004. J Glaciol Geocryol 29:351–359

Li Z, Yan Z (2009) Homogenized daily mean/maximum/minimum temperature series for China from 1960–2008. Atmos Oceanic Sci Lett 2:237–243

Li Q, Chen Y, Shen Y, Li X, Xu J (2011) Spatial and temporal trends of climate change in Xinjiang, China. J Geog Sci 21:1007–1018. https://doi.org/10.1007/s11442-011-0896-8

Long YL, Xia YL, Xia XY (2016) Temperature change characteristics and mutation analysis in Yanqi Basin from 1961 to 2014. Mod Agric Sci Technol

Luterbacher J, Dietrich D, Xoplaki E, Grosjean M, Wanner H (2004) European seasonal and annual temperature variability, trends, and extremes since 1500. Science 303:1499–1503

Ren G, Xu M, Chu Z, Guo J, Li Q, Liu X, Wang Y (2005) Changes of surface air temperature in China during 1951–2004. Clim Environ Res 10:717–727

Shi Y (2003) Discussion on the present climate change from warm-dry to warm wet in northwest china. Quat Sci 23:152–164

Solomon S (2007) Climate change 2007: the physical science basis: contribution of Working Group I to the Fourth Assessment Report of the Intergovernmental Panel on Climate Change. Cambridge University Press

Wang S, Zhu J, Cai J (2004) Interdecadal variability of temperature and precipitation in China since 1880. Adv Atmos Sci 21:307–313. https://doi.org/10.1007/bf02915560

Wang WX, Wang XJ, Jiang FQ, Peng DM (2012) Temperature and precipitation along the Kaidu River over the past 50 years. Arid Land Geogr 35:746–753

Xie Y, Zhu J (2011) Hydrological characteristics of Kaidou river basin. J Hydrol 31:92–96

You Q, Kang S, Aguilar E, Pepin N, Flügel W-A, Yan Y, Xu Y, Zhang Y, Huang J (2011) Changes in daily climate extremes in China and their connection to the large scale atmospheric circulation during 1961–2003. Clim Dyn 36:2399–2417

Zhang SM, Zhang KY, Xiao-Chuan LI, Zhou XY (2012) Temperature characteristics in winter at Yanqi Basin during 1951–2010 and its effects on agricultural production. Desert Oasis Meteorol

Zhen LI, Jiang FQ (2007) A study of abrupt climate change in Xinjiang Region during 1961–2004. J Glaciol Geocryol 29:351–359

Characteristics of Soil Organic Matter and Carbon and Nitrogen Contents in Crops/Plants: Land Use Impacts

Juan Zhang, Xiujun Wang, Jiaping Wang and Qingfeng Meng

Abstract Change of land use is an important factor for the dynamics of soil organic matter and carbon and nitrogen cycles. Many parts of native land in the Yanqi Basin have been converted to cropland since 1950. To study the influence of land use change on carbon and nitrogen cycling, we collected plant and soil samples from 28 representative sites in crop and native land. Our results show a narrow range in carbon content (i.e., 44.2% in the native species and 41.7% in the crop species), but a large range in nitrogen content (0.3–3.9%). Plant C:N ratio may follow an order: native species (~31) < crops (~53), and aboveground tissues (~32) < belowground tissues (~52). Soil C:N ratio in native land (9.9) is closed to that in cropland. Both soil organic carbon and total nitrogen show a marked increase after long-term cultivation in Yanqi Basin, which may be associated with the increase of recalcitrant SOC.

1 Introduction

Soil fertility is the basis for plant growth, and soil organic matter (SOM) is the basis for soil fertility. The quality of SOM can be characterized by carbon and nitrogen contents and fractionations of soil organic carbon (SOC). On the one hand, carbon (C) and nitrogen (N) contents in plants may reflect soil nutrient conditions; On the other hand, they can affect the characteristics of SOM. The arid and semiarid regions are a major component of globe biogeochemical cycle because they account for 1/3 earth's land surface. But less attention has been paid to the characteristics of SOM and C and N contents in the plants in the arid and semiarid regions.

J. Zhang (✉) · Q. Meng
School of Resources and Environment, Northeast Agricultural University, 150030 Harbin, China
e-mail: zhangjuan2080@163.com

X. Wang
College of Global Change and Earth System Science, Beijing Normal University, 100875 Beijing, China

J. Wang
College of Agriculture, Shihezi University, 832000 Shihezi, China

© Springer Nature Singapore Pte Ltd. 2018
X. Wang et al. (eds.), *Carbon Cycle in the Changing Arid Land of China*,
Springer Earth System Sciences, https://doi.org/10.1007/978-981-10-7022-8_4

The SOC dynamics reflects the equilibrium of input and output. Land use change as an important factor may change SOC content by altering the rates of litter inputs and/or SOM decomposition (Zhang et al. 2014b). In order to ensure food security, many parts of native land have been converted to cropland. Some studies conducted in tropical humid area have demonstrated that SOC content often declined after cultivation (Del Grosso et al. 2009; Dinesh et al. 2003). However, the opposite results were observed in arid regions. Fallahzade and Hajabbasi (2012) showed that content of total SOC was 5 times higher in cropland than that in native land; Another study also proved that labile carbon and semi-labile carbon increased after native land cultivation (Cochran et al. 2007).

Our study was conducted in Yanqi Basin that is a typical arid land. Many parts of native land have been used for cropping since 1950 in this region. We determine C and N contents in typical vegetations and soils from native and cropland. Our study aims to assess C and N distributions in plants and soils, and to determine the relationship among them.

2 Materials and Methods

2.1 Site Description

Our sampling area is located in the center of Yanqi Basin, which is a typical arid land in Northwest China (Fig. 1). This region shows annual average temperature <10 °C, with a very low precipitation (<80 mm) but very high evaporation (>2000 mm). The major soil types are brown/gray-brown desert soil. The typical native plants are *Halostachys caspica C. A. Mey. ex Schrenk* and *Phragmites australis (Cav.) Trin. ex Steud.* But portions of native land have been cultivated after 1950. *Lycopersicon esculentum Mill.* and *Capsicum annuum Linn* are the dominant crops.

2.2 Plant and Soil Sampling and Analysis

We collected plant and soil samples randomly from 14 sites in native land and 14 sites in cropland in fall, 2010. Plant samples were gathered from two parts: aboveground tissue and root, which were grinded to 0.15 mm for C and N contents measurements, by using an element analyzer.

Soil samples were gathered from five layers, which were air-dried then passed 2 mm/0.25 mm sieve. Soil pH was measured using 2-mm soils with 1:5 soil-to-water ratio. SOC and TN were analyzed using the 0.25-mm soils, by following the Walkley–Black method (Walkley and Black 1934), and fully automatic azotometer, respectively. For the measurement of SOC fractions, we followed a two-step H_2SO_4

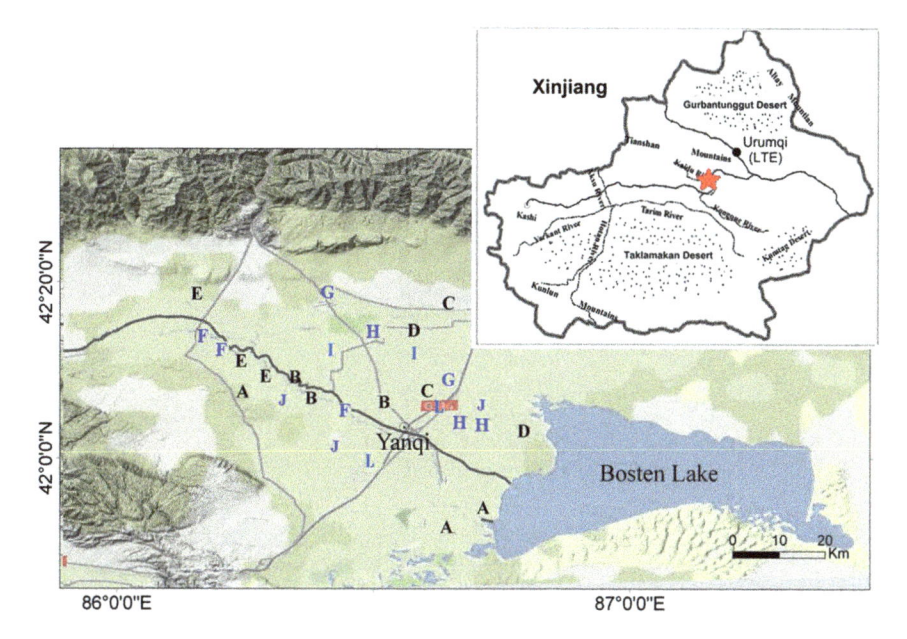

Fig. 1 Sampling sites (modified from Zhang et al. 2014b)

hydrolysis procedure (Rovira and Vallejo 2002). Plant species and soil properties in surface layer are shown in Table 1.

3 Results and Discussion

3.1 Plant C Contents

Plant C content in native and crop species is shown in Fig. 2. A wide range of C content is seen for native species, i.e., 38.5–49.5% for aboveground tissues, and 37.8–50.1% for roots (Fig. 2a). On the other hand, a relatively smaller range of C content (37.1–44.1%) is found in crop species for the whole plants (Fig. 2b). On average, C content in native species (44.2%) is close to that in crop species (41.7%).

Relative to other studies, C content is modestly higher in the crops/plants of the Yanqi Basin. For example, Xu et al. (2007) and Dossa et al. (2009) showed that the C content was about 35–36% in shrub species in some other arid regions, i.e., the Inner Mongolia and Senegal. The lower C content in those places might be associated with the extremely low levels of SOC (~4 g kg^{-1}).

Table 1 Characteristics of plant and soil among sampling sites (modified from Zhang et al. 2014a, b)

Land use types	Sites	Plant species	BD (g cm^{-3})	pH
Native land	A	*Phragmites australis (Cav.) Trin. ex Steud.*	1.5	8.1
	B	*Alhagi sparsifolia Shap.*	1.4	8.5
	C	*Tamarix ramosissima Ledeb.*	1.6	8.6
	D	*Halostachys caspica C. A. Mey. ex Schrenk*	1.2	8.5
	E	*Populus tomentosa Carr/Acroption repens DC. Prodr./Sophora alopecuroides Linn*	1.5	8.1
Cropland	F	*Capsicum annuum Linn*	1.4	8.0
	G	*Solanum lycopersicum*	1.3	8.4
	H	*Zea mays*	1.3	8.3
	I	*Lycopersicon esculentum Mill.*	1.3	8.4
	J	*Gossypium spp/Brassica campestris L./Helianthus annuus/Beta vulgaris*	1.3	8.3

Note BD bulk density

3.2 Plant N Contents

The N content reveals great variations between different plant tissues and among sites (Fig. 3). For native plants, N content varies from 1.3 to 3.9% for above and from 0.7 to 2.0% for roots. On average, the N content in aboveground (2.5%) is almost twice of that in belowground tissues (1.3%). For crops, a narrow range of N content is observed, with 0.8–1.3% for aboveground and 0.3–1.3% for roots. The mean N content is similar (~1.0%) for different parts in crops.

Fig. 2 Plant C content in native (**a**) and crop (**b**) species

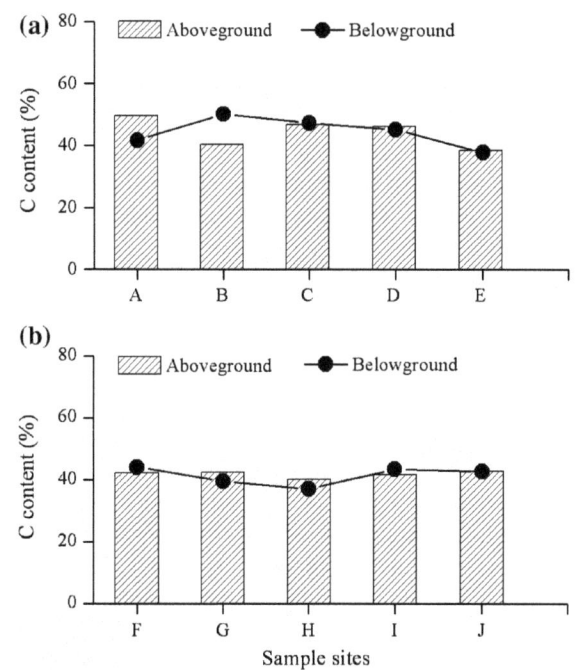

Fig. 3 Plant N content in native (**a**) and crop (**b**) species

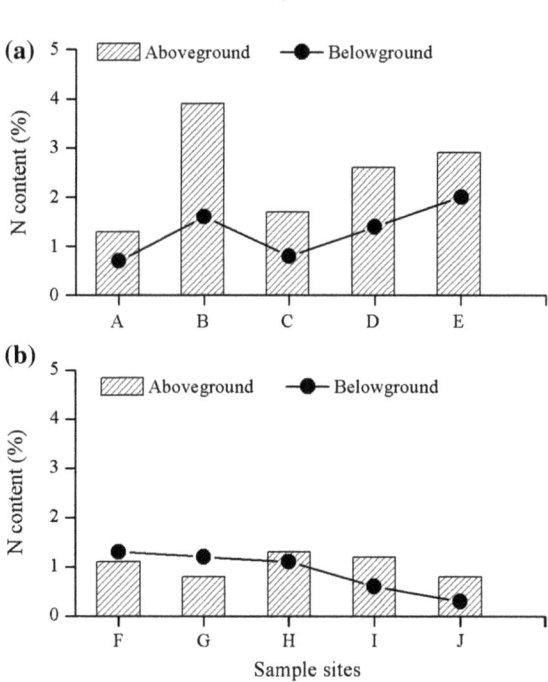

Early studies revealed that N content in shrub was between 2.3–2.6 for aboveground and 2.8 for belowground tissues (Xu et al. 2007; Yuan et al. 2005); comparing with those in the Yanqi Basin, N content was comparable for aboveground part, but higher in belowground part. As for crops, Zhai et al. (2013) reported a large range (0.4–3.3%) for wheat N content in Henan Province of China. Minkenberg and Ottenheim (1990) illustrated that plant N content ranged from 2.1 to 4.9% in the Western Europe. Apparently, nitrogen content is significantly lower in the crops of Yanqi Basin than in those of other parts in the world, probably owing to the low soil fertility.

3.3 Soil C and N

The SOC and TN show almost identical vertical distribution in the native land (Fig. 4). For most sites, the SOC is <10 g kg^{-1} and TN < 1 g kg^{-1} in topsoil layer; both SOC and TN show a clear decline with depth above 40 cm. Similar results were also reported for northern Xinjiang by Li et al. (2010).

Figure 5 shows that there are some differences in the vertical distribution between SOC and TN in the cropland although both show a sharp decrease with depth. Apparently, the cropland sites contain much higher levels of SOC and TN relative to the native land. For example, SOC content is almost greater than 10 g kg^{-1} and TN content greater than 1 g kg^{-1} in the top three soil layers. There has been evidence that land reclamation in arid regions often leads to an increase in SOC and TN contents, as reported for northern Xinjiang (Li et al. 2010) and Inner Mongolia (Wang et al. 2010).

3.4 Soil Organic Carbon Fractions

3.4.1 SOC Fractions and SOC Stock in the Native Land

Generally, C content in each fraction is higher in topsoil layer than in deeper layers (see Fig. 6). For example, labile SOC is always >1 g kg^{-1} for the top 10 cm, but <1 g kg^{-1} in the other soil layers. And the content of recalcitrant SOC has a range of 3.42–5.73 g kg^{-1} for the surface layer, but <2 g kg^{-1} below 50 cm. The content of semi-labile SOC is lowest among the SOC fractions, i.e., <1 g kg^{-1} for the whole profiles. The SOC stock in Yanqi Basin is 6.01 kg m^{-2}, which composed of 1.34 kg m^{-2} from labile SOC, 0.89 kg m^{-2} from semi-labile SOC, and 3.78 kg m^{-2} from recalcitrant SOC (Table 2).

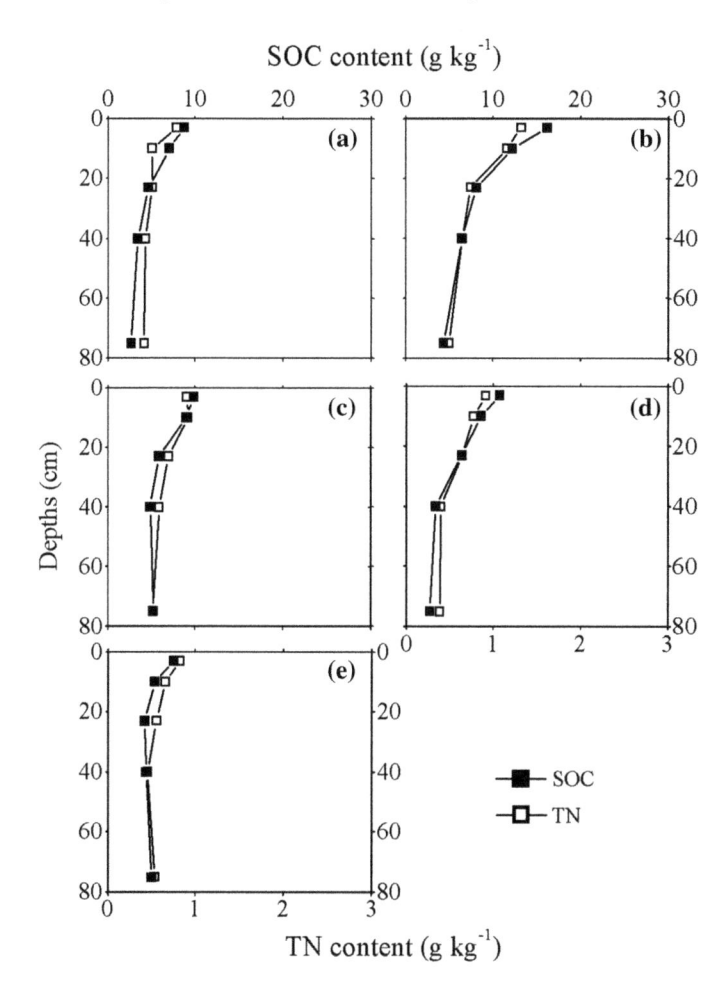

Fig. 4 Content of SOC and TN in the native land (modified from Zhang et al. 2014a)

Table 2 SOC stock in each fraction (after Zhang et al. 2014b)

Land use types	Labile carbon	Semi-labile carbon	Recalcitrant carbon
Native land (kg m^{-2})	1.34	0.89	3.78
Cropland (kg m^{-2})	2.31	1.37	6.50
Difference	0.97	0.48	2.72

3.4.2 SOC Fractions and SOC Stock in the Cropland

All SOC fractions in cropland are significantly higher relative to those in the native land (Fig. 7). The SOC content of each fraction is generally higher in top three layers in the cropland, but shows a sharp decrease over the 30–100 cm. For example, the

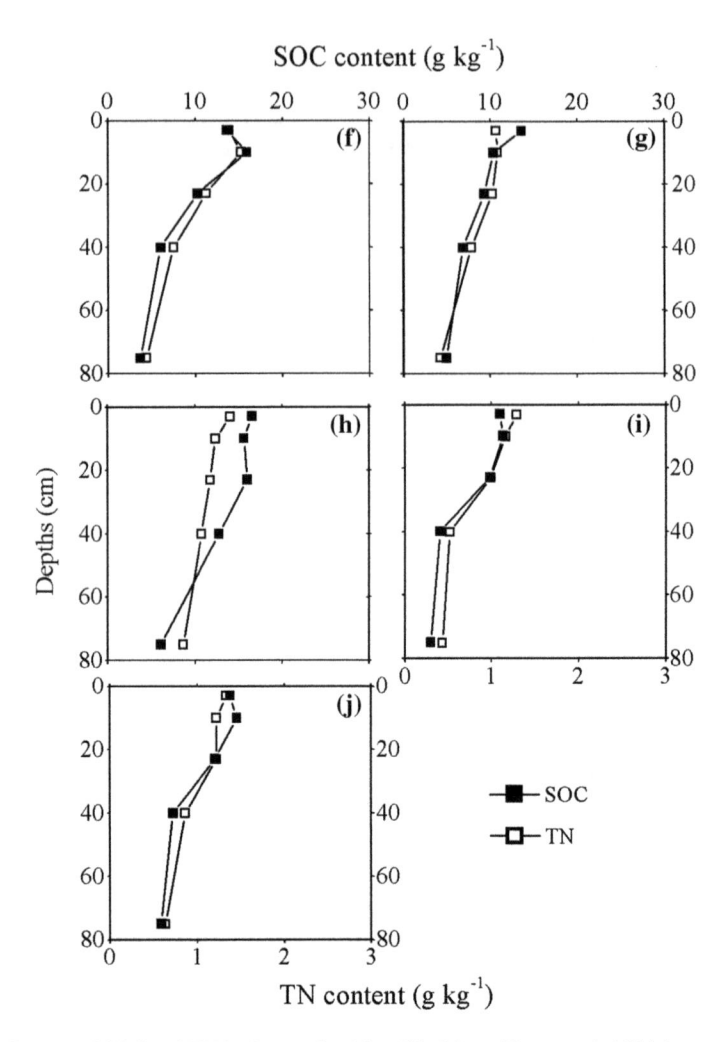

Fig. 5 Content of SOC and TN in the cropland (modified from Zhang et al. 2014a)

labile SOC is 2.25–2.02 g kg^{-1} for the top 30 cm, but lower than 1.5 g kg^{-1} at the depth of 30–100 cm. The semi-labile carbon is higher than 1 g kg^{-1} for the surface 30 cm, but <0.7 g kg^{-1} in the 30–100 cm. The recalcitrant SOC is almost >8 g kg^{-1} in top three layers, but <3.5 g kg^{-1} in the deeper layers.

After long-term cultivation in Yanqi Basin, the SOC stock increased 4.17 kg m^{-2} (Table 2). The study by Zhang et al. (2012) showed that SOC stock increased by 1.2 kg m^{-2} after 20 years of tillage. The significant increase of SOC stock in cropland may be associated with agricultural practice. For example, manure application and residue management can also lead to SOC enhancement in cropland (Tian et al. 2015; Benbi et al. 2015).

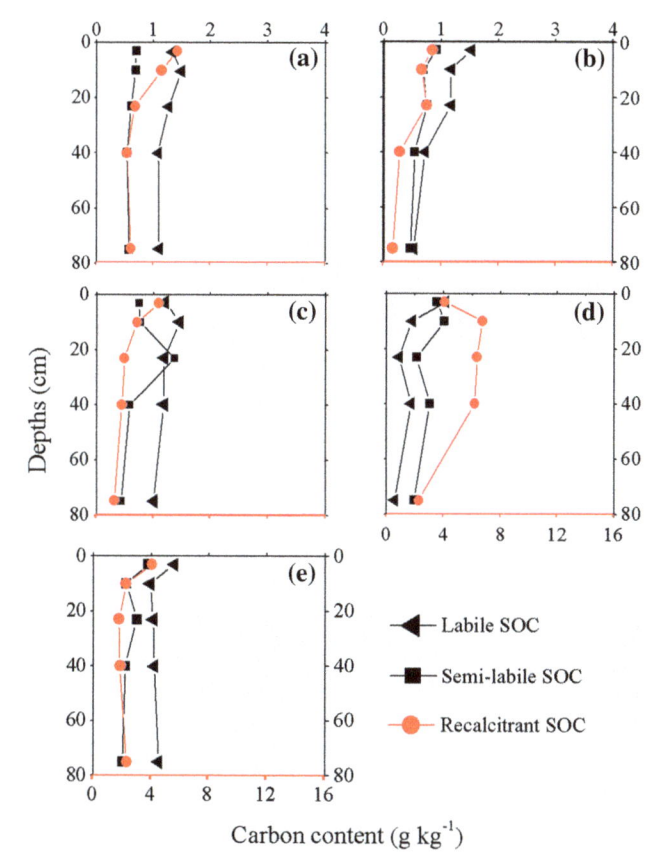

Fig. 6 SOC fractions in the native land (modified from Zhang et al. 2014b)

The contribution of each fraction to the increase in SOC stock is different, i.e., 0.97, 0.48, and 2.72 kg m^{-2} from labile, semi-labile, and recalcitrant SOC, respectively. Obviously, recalcitrant SOC has a primary role to enhance SOC stock in the cropland. Given that recalcitrant SOC has a direct relationship with long-term carbon storage, one could suggest that cultivation may be conducive to long-term carbon storage in the arid land (Zhang et al. 2014b).

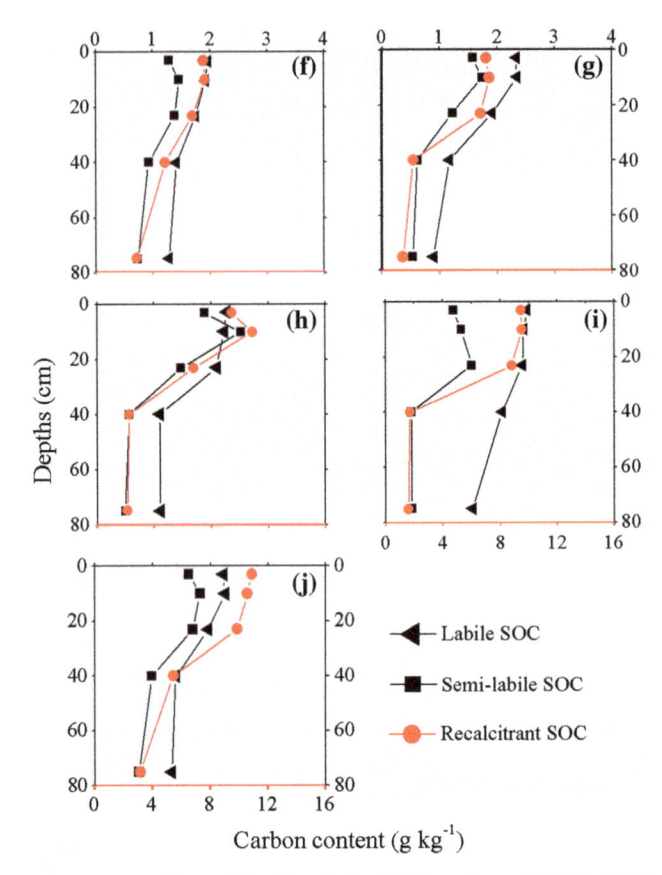

Fig. 7 SOC fractions in the cropland (modified from Zhang et al. 2014b)

3.5 C:N Ratio in Soil-Plant Systems

3.5.1 Plant C:N Ratio

Plant C:N ratio can have effects on decomposition of litters. In general, the higher C:N ratio of plant, the more difficult the decomposition is (Jafari et al. 2011). As illustrated in Fig. 8, plant C:N ratio is generally lower in the native land, with an average of 21.4 for the aboveground parts and 40.2 for the roots. Similar results were obtained by Mi et al. (2005), revealing extremely higher C:N ratio in roots for arid plant. The C:N ratio is much higher in the crops than in the native plants, with 42.2 for aboveground and 63.1 for belowground tissues. These results suggested that the decomposition

Fig. 8 Plant C:N ratio in the native (a) and crop (b) species

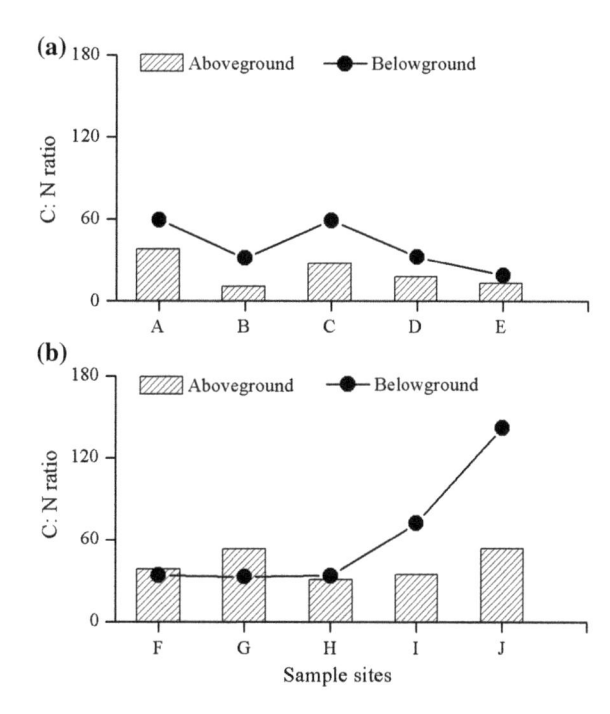

Table 3 Soil C:N ratio in the native land and cropland

Depths (cm)	Native land	Cropland
0–5	11.5	10.7
5–15	11.0	10.9
15–30	10.4	10.3
30–50	8.3	9.0
50–100	8.6	8.6

rate may follow an order: native species > crops, and aboveground > belowground tissues.

3.5.2 Soil C:N Ratio

On average, soil C:N ratio in native land (9.9) is closed to that in cropland (10.0, see Table 3). But the C:N ratio in top three layers (10.8) is higher than in the 30–100 cm (8.6). Soil C:N ratio is usually affected by various factors, i.e., manure application, vegetation types, and microbial activities (Esser et al. 2011). There was evidence of a negative correlation between SOM decomposition rate and soil C:N ratio (Zhang et al. 2008; Wang et al. 2015).

4 Concluding Remarks

Knowledge of C and N distributions in vegetation and soil is significant for comprehending the biogeochemical cycle on the arid land. Our study shows that C content in the native species (44.2%) is close to that in the crops (41.7%), but N content varies largely (0.3–3.9%). Plant C:N ratio follows an order: native species < crops, and aboveground tissues < belowground tissues. Soil C:N ratio in native land (9.9) is closed to that in cropland in Yanqi Basin.

Both SOC and TN show a marked increase after intensive cropping in Yanqi Basin, which may be associated with the increase of recalcitrant SOC fraction. Future work should include the studies of plant–soil interactions (e.g., allocation of organic carbon), the interactions and feedbacks between the carbon cycle and nitrogen cycle, and the effect of land use change and climate change in the arid region.

References

Benbi DK, Brar K, Toor AS, Sharma S (2015) Sensitivity of labile soil organic carbon pools to long-term fertilizer, straw and manure management in rice-wheat system. Pedosphere 25:534–545

Cochran R, Collins H, Kennedy A, Bezdicek D (2007) Soil carbon pools and fluxes after land conversion in a semiarid shrub-steppe ecosystem. Biol Fertil Soils 43:479–489

Del Grosso SJ, Ojima DS, Parton WJ, Stehfest E, Heistermann M, DeAngelo B, Rose S (2009) Global scale DAYCENT model analysis of greenhouse gas emissions and mitigation strategies for cropped soils. Glob Planet Change 67:44–50

Dinesh R, Chaudhuri SG, Ganeshamurthy AN, Dey C (2003) Changes in soil microbial indices and their relationships following deforestation and cultivation in wet tropical forests. Appl Soil Ecol 24:17–26

Dossa E, Khouma M, Diedhiou I, Sene M, Kizito F (2009) Carbon, nitrogen and phosphorus mineralization potential of semiarid Sahelian soils amended with native shrub residues. Geoderma 148:251–260

Esser G, Kattge J, Sakalli A (2011) Feedback of carbon and nitrogen cycles enhances carbon sequestration in the terrestrial biosphere. Glob Change Biol 17(2):819–842

Fallahzade J, Hajabbasi M (2012) The effects of irrigation and cultivation on the quality of desert soil in central Iran. Land Degrad Dev 23:53–61

Jafari M, Kohandel A, Baghbani S, Tavili A, Chahouki MAZ (2011) Comparison of chemical characteristics of shoot, root and litter in three range species of Salsola rigida, Artemisia sieberi and Stipa barbata. Caspian J Environ Sci 9:37–46

Li C, Li Y, Tang L (2010) Soil organic carbon stock and carbon efflux in deep soils of desert and oasis. Environ Earth Sci 60:549–557

Mi HL, Xu X, Li SH, He J, Ma YM (2005) Dynamic change of the contents, distributions and ratioes of carbohydrate and total nitrogen in Cynanchum komarovii and Glycyrrhiza uralensis during the different periods of growth. Agric Res Arid Areas 23:129–133 (in Chinese)

Minkenberg OPJM, Ottenheim JJGW (1990) Effect of leaf nitrogen content of tomato plants on preference and performance of a leafmining fly. Oecologia 83:291–298

Rovira P, Vallejo VR (2002) Labile and recalcitrant pools of carbon and nitrogen in organic matter decomposing at different depths in soil: an acid hydrolysis approach. Geoderma 107:109–141

Tian K, Zhao YC, Xu XH, Hai N, Huang B, Deng WJ (2015) Effects of long-term fertilization and residue management on soil organic carbon changes in paddy soils of China: a meta-analysis. Agric Ecosyst Environ 204:40–50

Walkley A, Black IA (1934) Estimation of soil organic carbon by the chromic acid titration method. Soil Sci 37:29–38

Wang Y, Li Y, Ye X, Chu Y, Wang X (2010) Profile storage of organic/inorganic carbon in soil: from forest to desert. Sci Total Environ 408:1925–1931

Wang YH, Gong JL, Liu M (2015) Effects of land use and precipitation on above- and below-ground litter decomposition in a semi-arid temperate steppe in Inner Mongolia, China. Appl Soil Ecol 96:183–191

Xu Z, Zhou G, Wang Y (2007) Combined effects of elevated CO_2 and soil drought on carbon and nitrogen allocation of the desert shrub Caragana intermedia. Plant Soil 301:87–97

Yuan ZY, Li LH, Huang JH, Jiang GM, Wan SQ, Zhang WH, Chen QS (2005) Nitrogen resorption from senescing leaves in 28 plant species in a semi-arid region of northern China. J Arid Environ 63:191–202

Zhai QY, Zhang J, Xiong SP (2013) Research on hyperspectral differences and monitoring model of leaf nitrogen content in wheat based on different soil textures. Sci Agricultura Sin 46(13):2655–2667

Zhang DQ, Hui DF, Luo YQ (2008) Rates of litter decomposition in terrestrial ecosystems: global patterns and controlling factors. J Plant Ecol 1(2):85–93

Zhang J, Wang XJ, Wang JP (2014a) Impact of land use change on profile distributions of soil organic carbon fractions in the Yanqi Basin. CATENA 115:79–84

Zhang J, Wang XJ, Wang JP, Wang WX (2014b) Carbon and nitrogen contents in typical plants and soil profiles in Yanqi Basin of Northwest China. J Integr Agric 13(3):648–656

Zhang L, Xie Z, Zhao R, Wang Y (2012) The impact of land use change on soil organic carbon and labile organic carbon stocks in the Longzhong region of Loess Plateau. J Arid Land 4:241–250

Dynamics of Soil CO_2 and CO_2 Efflux in Arid Soil

Junyi Wang, Xiujun Wang, Jiaping Wang and Tongping Lu

Abstract Soil carbon dioxide (CO_2) is an important component in the terrestrial ecosystem and regulates the atmosphere–land CO_2 exchange. Studying the variation of CO_2 in soil profile and surface CO_2 efflux can improve our understanding of the carbon cycle in the terrestrial ecosystem. In this paper, we analyze the dynamics of soil CO_2 and CO_2 efflux in an arid region, Yanqi, northwest China. Both CO_2 concentration and surface CO_2 efflux showed a clear seasonal variation, with two peaks in summer and a gradual decrease in autumn. We found that surface CO_2 efflux was exponentially related to soil temperature and linearly related to moisture when soil moisture was less than ~22%. We estimated surface CO_2 efflux by linear regression of CO_2 fluxes that were calculated by gradient method using Penman (1940), Marshall (1959), and Moldrup et al. (2013) models and found that the Marshall (1959) model did a better job than the other two models. However, there were considerable mismatches between the observation and model results. Our study indicates that the relationship is complex between the concentration of CO_2 in soil profile and surface CO_2 efflux in the arid region.

1 Introduction

Soil carbon dioxide (CO_2) is an important component in the terrestrial ecosystem and regulates the atmosphere–land CO_2 exchange. Recently, some studies (Jia et al. 2014; Li et al. 2015; Liu et al. 2015; Xie et al. 2009) showed a great potential of CO_2 uptake in arid and semiarid regions of the northwest China. For example, Xie et al. (2009) reported that CO_2 uptake in the desert soils could reach 62–622 g C m^{-2} $year^{-1}$. Therefore, studying the dynamics of soil CO_2 and CO_2 diffusion mechanisms in

---

J. Wang · X. Wang (✉) · T. Lu
College of Global Change and Earth System Science, Beijing Normal University, 100875 Beijing, China
e-mail: xwang@bnu.edu.cn

J. Wang
College of Agriculture, Shihezi University, 832000 Shihezi, China

© Springer Nature Singapore Pte Ltd. 2018
X. Wang et al. (eds.), *Carbon Cycle in the Changing Arid Land of China*,
Springer Earth System Sciences, https://doi.org/10.1007/978-981-10-7022-8_5

arid region will improve our understanding of the carbon cycle in the terrestrial ecosystems.

Soil CO_2 concentration depends on many factors, including soil moisture, temperature, organic matter content, and other soil properties, without considering the vegetation. There were studies on soil CO_2 concentration using sampling tubes (Fierer et al. 2005; Kammann et al. 2001; Sanderman and Amundson 2010), which cannot provide continuous measurement or minimize disturbance of soil environment. However, the employment of solid-state CO_2 sensors could resolve these issues. For example, Tang et al. (2003) carried out a study in a Mediterranean savanna ecosystem of California and validated the feasibility of CO_2 sensors.

At present, the common methods for determining surface CO_2 efflux are eddy covariance method (Pattey et al. 2002), chamber-based method (Wang et al. 2010), and gradient method (Tang et al. 2003). The first two methods only provide CO_2 exchange between the ecosystem/soil and atmosphere; however, gradient method can provide more information about CO_2 dynamics in soil profile, including CO_2 diffusion at different depths. However, gradient method may introduce biases and uncertainties due to some assumptions dealing with gas diffusivity (Risk et al. 2015), spatial heterogeneity (Wiaux et al. 2015), and the interferences by other factors such as wind (Maier et al. 2010) and rainfall (Tang et al. 2005).

Although there have been some studies on surface CO_2 efflux in China's arid area (Liu et al. 2017; Yan et al. 2014; Zhang et al. 2010), our understanding of soil CO_2 dynamics and its relationship with CO_2 efflux is limited. Therefore, we carried out an integrative study during the summer and autumn of 2012 in Yanqi Basin, Xinjiang, by continuously monitoring CO_2 concentration, temperature, and water content in soil profile and simultaneously measuring surface CO_2 efflux. The objectives of our study are as follows: (a) to evaluate the temporal and vertical variations of CO_2 concentration, (b) to analyze the temporal variability of soil CO_2 efflux and influence of environmental factors, and (c) to evaluate the applicability of different diffusion models for the arid soil.

2 Materials and Methods

2.1 Measurement of Soil CO_2, Water Content, and Temperature

Straw return is an effective practice for sustaining crop productivity and soil fertility in large parts of China (Wang et al. 2015b). Straw returning is a common agricultural management in Xinjiang. Accordingly, the experiment included 2 PVC tubes (50 cm in diameter and 80 cm in height), corresponding two treatments: soil only and soil mixed with maize straw (<2 cm in length). Before adding soil, the two tubes were buried underground, spacing 50 cm, the top as high as the ground level. We collected surface soil (0–30 cm) after harvest, which was passing through a 5-mm sieve (with root removed), and fully mixed and added water so that soil moisture content was close to 60% of field capacity. Then, we selected randomly two samples to be used in experiment and selected two representative soil samples to determine the physical and chemical properties. The soil is a silty loam composed of 8.5% clay, 72.7% silt, and 18.8% sand, which contains organic carbon of 1.09%, inorganic carbon of 2.29%, total nitrogen of 0.09%, with pH of 8.1 and a bulk density of 1.3 g cm^{-3}.

We used CO_2 sensors (GMT222, Vaisala Inc., Finland) to monitor CO_2 concentration in soil profiles. To avoid the damage of water to probes, before installation, all CO_2 sensors were housed inside steel tubes, the upper end sealed with plastic film, and the lower end wrapped with waterproof-breathable membrane (Gore-Tex). We vertically put one CO_2 sensor just above the soil surface and buried two CO_2 sensors into soil at depths of 10 and 20 cm in each PVC.

Soil temperature (T_S) and soil water content (SWC) were measured using temperature probe (109, Campbell Scientific Inc.) and water content reflectometers (CS616, Campbell Scientific Inc.) with precision of better than 0.1% volumetric water content. These probes were placed horizontally at the same depths where the CO_2 sensors were installed.

Because the soil sank during the experiment, we remeasured the depths of all probes in the soil at the end of the experiment, which was 6, 17 cm in the soil-only treatment and 8, 18 cm in the soil with straw treatment, respectively. All outputs from sensors were scanned every 30 min, and hourly means were computed and stored in the data logger (CR1000, Campbell Scientific Inc.). Data for analysis were collected from June 29, 2012, to November 19, 2012.

2.2 Soil CO_2 Efflux Measurements

In this experiment, we measured the soil surface CO_2 efflux using a LI-8150 CO_2 efflux system (LI-COR, USA). Two soil collars with a height of 20 cm and a diameter of 21.3 cm were pushed into the soil for 10 cm. One long-term measurement chamber was installed in each soil collar so that temporal variation of CO_2 effluxes of two treatments can be independently observed. During the study period, the LI-8150 automatically recorded every half hour of CO_2 efflux.

2.3 Calculation of CO_2 Diffusion

Molecular diffusion is the dominant process of CO_2 transport in the soil profile and soil–atmosphere boundary. Accordingly, we can use gradient method (Jone and Schapper 1972) based on Fick's first law to calculate CO_2 flux of each soil layer:

$$F = -D_S \frac{dC}{dz} \quad (1)$$

where F is the CO_2 flux (μmol m^{-2} s^{-1}), D_S the CO_2 diffusion coefficient (m^2 s^{-1}), C the CO_2 concentration (μmol m^{-3}), and $\frac{dC}{dz}$ the vertical gradient of soil CO_2.

D_S can be derived by a function of the diffusivity of CO_2 in free air (D_a) and the gas tortuosity factor (ξ) as follow:

$$D_S = \xi D_a \tag{2}$$

We compute ξ using three models, according to Penman (1940), Marshall (1959), and Moldrup et al. (2013), respectively:

$$\xi = 0.66\varepsilon \tag{3}$$

$$\xi = \varepsilon^{1.5} \tag{4}$$

$$\xi = \varepsilon^{[1+C_m\emptyset]}\left(\frac{\varepsilon}{\emptyset}\right) \tag{5}$$

where ε is air-filled porosity in soil, $\emptyset$ the total porosity, and C_m the media complexity factor, with $C_m = 1$ in repacked soils.

The influence of temperature on the diffusion coefficients is given by Tang et al. (2003):

$$D_a = D_{a0}\left(\frac{T}{293.15}\right)^{1.75} \tag{6}$$

where T is the temperature (K) and D_{a0} the standard diffusion coefficient at 20 °C (293.15 K) and 101.3 kPa, which is 1.47×10^{-5} m^2 s^{-1} (Jones 1992).

2.4 Computation of the Relationships of CO_2 Efflux with Soil Temperature and Moisture

We used an exponential relationship and a quadratic relationship to fit soil CO_2 efflux against temperature (T) and moisture (M), respectively:

$$F = ae^{bT} \tag{7}$$

$$F = a + bM + cM^2 \tag{8}$$

where a, b, and c are fitted parameters.

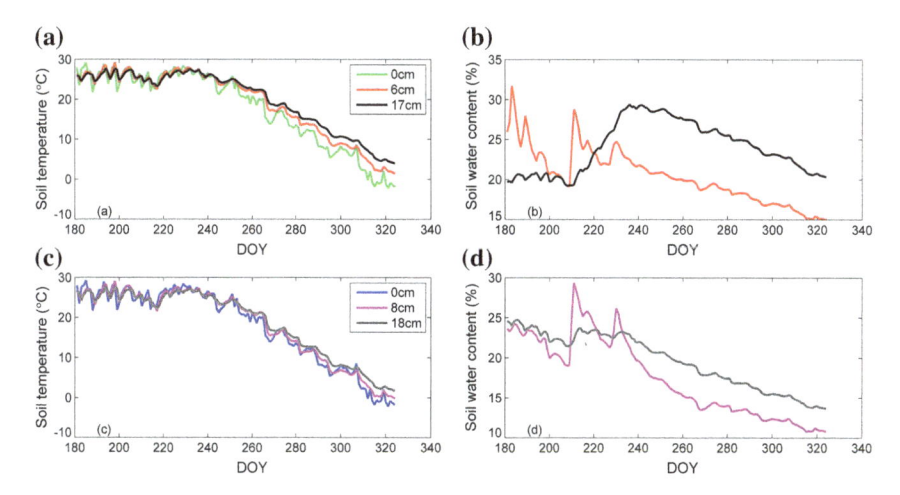

Fig. 1 Seasonal variation of (**a, c**) soil temperature and (**b, d**) soil water content at different depths in the soil-only treatment (upper panel) and soil with straw treatment (lower panel) (left panel redrawn from Wang et al. 2015a)

3 Results

3.1 Seasonal Variations of Soil Variables

Figure 1 shows seasonal variations in T_S and SWC at different soil depths. During the observation period, T_S showed a similar pattern in the soil-only treatment (Fig. 1a) and soil with straw treatment (Fig. 1c). Daily mean T_S remained relative high in summer with small fluctuation from 21 to 30 °C, while T_S in autumn gradually decreased from 26 to −3 °C. Seasonal fluctuation of T_S was greater near the surface than in the deep layers. For example, T_S ranged from 29.3 to −2.2 °C (in the soil-only treatment) and 29.3 to −2.2 °C (in the soil with straw treatment) at 0 cm and ranged from 27.8 to 3.9 °C (in the soil-only treatment) and 27.7 to 1.7 °C (in the soil with straw treatment) at 17 and 18 cm.

The day-to-day variation of SWC differed substantially between the soil only and soil with straw profiles. In the soil-only treatment (Fig. 1b), SWC at 6 cm depth gradually decreased in the mid-summer and reached two peaks (28.8 and 24.8%, respectively) in late summer due to two significant rainfall events. However, SWC at 17 cm depth remained relatively stable in the mid-summer and then continuously increased in late summer and reached its maximum (29.4%) after second rainfall event. In autumn, SWC showed a similar decreasing trend at two depths. However, SWC in the soil with straw treatment (Fig. 1d) at 18 cm depth generally displayed a decrease tendency with weak increase caused by two rainfall events. The evolution of SWC at 8 cm in the soil with straw treatment resembled that at 7 cm in the soil-only treatment with peaks (29.4 and 26.2%, respectively) after rainfall.

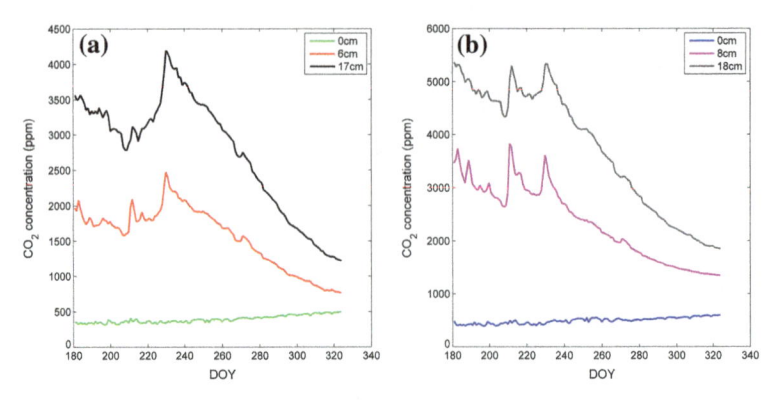

Fig. 2 Seasonal variation of soil CO_2 concentration at different depths in the soil-only treatment (**a**) and soil with straw treatment (**b**) between day of 181 and 324 in 2012 (after Wang et al. 2015a)

Seasonal variations in soil CO_2 concentration in the two treatments are shown in Fig. 2. During the study period, CO_2 concentration at soil surface generally kept steady and slightly increased from 355.1 to 495.1 ppm in the soil-only treatment (Fig. 2a) and from 388.7 to 596.3 ppm in the soil with straw treatment (Fig. 2b). In the soil-only treatment, CO_2 concentration averaged 1558 and 2753 ppm at 6 and 17 cm, respectively, which were lower than the CO_2 concentration in the soil with straw treatment, which averaged 2346 and 3720 ppm at the depth of 8 and 18 cm, respectively. In deeper soil, CO_2 concentration had an obvious seasonal variation, which dropped in late June and July, reached two peaks following the two rainfall events in August, and then continuously declined from September to November.

3.2 Relationship Between Observed CO_2 Efflux and Estimated CO_2 Flux

Figure 3 shows the relationship between observed surface CO_2 efflux and calculated CO_2 flux near the surface using three different diffusivity models. We selected 5 days (day 201–206 in 2012) in the mid-summer and 5 days (day 275–280 in 2012) in the mid-autumn to compare the seasonal difference in diurnal pattern of surface CO_2 efflux and magnitudes between observed and calculated values. Surface CO_2 effluxes in both treatments showed a similar diurnal variation with peaks at 13:00–17:00 and valleys at 22:00–08:00 in the mid-summer and mid-autumn. Three different diffusion models estimated fluxes at the top of the soil profiles with a similar diurnal variation, however, whose peaks and valleys lagged behind observed surface CO_2 effluxes. The values calculated by Penman (1940) model were always highest and those by Moldrup et al. (2013) model lowest. In the mid-summer, we found that calculated average CO_2 fluxes at the top of profiles by Penman (1940), Marshall (1959), and Moldrup et al. (2013) models were lower than average observed surface CO_2 effluxes

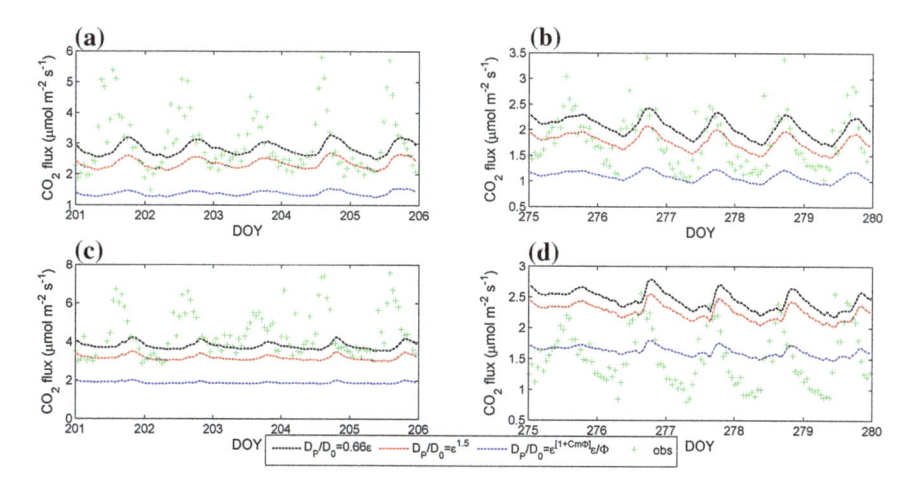

Fig. 3 Comparison of soil CO_2 effluxes between observed and calculated using three different diffusivity models in the soil-only treatment at 3 cm (**a**), (**b**) and soil with straw treatment at 4 cm (**c**), (**d**)

by 5.6–14%, 21–28%, 54–57%, respectively. In contrast, in the mid-autumn, we found that estimated mean values using the Penman (1940), Marshall (1959), and Moldrup et al. (2013) models were 62.1, 49.1, and 6.8% higher in the soil with straw treatment; in the soil-only treatment, the values estimated by Penman (1940) model were 14.7% higher; however, the Marshall (1959) and Moldrup et al. (2013) models underestimated the CO_2 efflux by 1.8 and 39.2%, respectively.

3.3 Vertical Profiles of CO_2 Fluxes

Figure 4 shows means of measured surface CO_2 efflux and CO_2 fluxes (at two depths) calculated using the concentration gradient method with three different diffusion models between day 201 and 205 in the mid-summer (Fig. 4a, c) and day 275–280 in the mid-autumn (Fig. 4b, d). Generally, the vertical profiles of calculated soil CO_2 fluxes showed a similar shape, i.e., soil CO_2 flux decreasing with depth. In the mid-summer, estimated CO_2 fluxes at two depths were less than the observed surface CO_2 effluxes in both treatments. However, in the mid-autumn, the calculated CO_2 flux at depth was higher than the measured surface CO_2 efflux. For example, in soil with straw treatment, the calculated CO_2 flux at 4 cm was higher relative to the measured surface CO_2 efflux.

One may estimate the surface CO_2 efflux using linear extrapolated method. However, using different models could result in dramatically different values. The Moldrup et al. (2013) model underestimated the surface CO_2 effluxes in the soil-only treatment, while Penman (1940) model overestimated the values for both treatments.

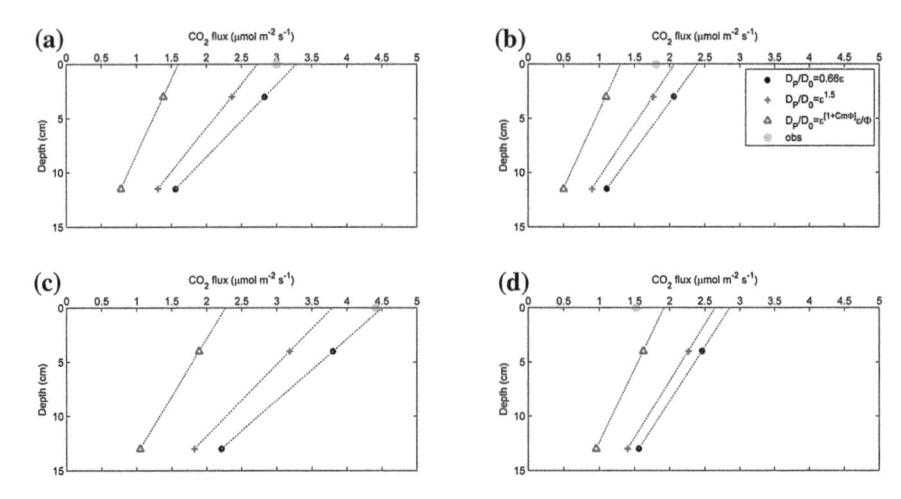

Fig. 4 Distribution of modeled average soil CO_2 fluxes and observed average surface CO_2 effluxes in (**a, b**) soil-only treatment and (**c, d**) soil with straw treatment between day 201 and 205 (left panel) and day 275–280 (right panel). The gray dashed lines represent linear regressions

Overall, the Marshall (1959) model produced smallest bias, i.e., overestimation of 4.7%.

3.4 Seasonal Variations of CO_2 Fluxes

Based on the evaluations of the three models, we found that the performance of the Marshall (1959) model was better than the other two. Thus, we used the Marshall model to calculate CO_2 fluxes in soil profiles. Figure 5 shows the measured daily mean surface CO_2 effluxes and estimated daily mean CO_2 fluxes over the entire experiment period. For both treatments, there was a strong seasonal variation, followed a bimodal curve with two peaks in July and August. Interestingly, the calculated CO_2 fluxes showed an immediate decrease after first rainfall episode, but a small increase near the surface following second rainfall event. These different responses were partly due to the intensity of the first rainfall greater than second times and partly due to a significant increase of CO_2 concentration gradient which offset the decrease of CO_2 diffusivity following the second rainfall.

We also assessed the effect of temperature and water content on surface CO_2 efflux using Eqs. 7–8 for the experiment period. Our analyses showed that soil CO_2 efflux was significantly ($P < 0.001$) correlated with temperature and water content in both treatments. More precisely, soil CO_2 efflux had a positive relationship with temperature (as an exponential function). However, the fitting relationship between CO_2 efflux and soil moisture was complex: positive when SWC < 24% (SWC < 22%) and negative when SWC > 24% (SWC > 22%) in the soil-only treatment (in the soil

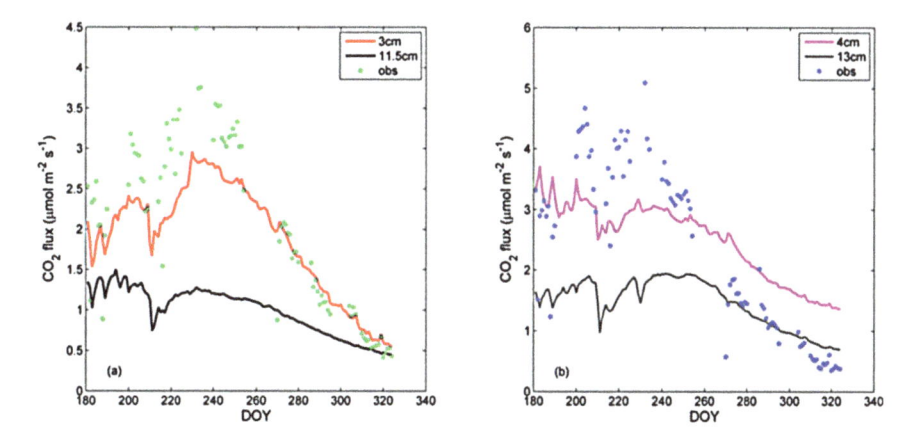

Fig. 5 Time series of daily mean soil CO_2 effluxes observed and CO_2 fluxes calculated for different depths using the Marshall (1959) diffusivity model in the soil-only treatment (**a**) and soil with straw treatment (**b**)

with straw treatment) as shown in Fig. 6. Nevertheless, it seems clear that there is a linear relationship between CO_2 efflux and soil moisture when soil water content is less than ~22%.

4 Discussion

4.1 The Effect of Environmental Factors on Soil CO_2 Efflux

There is evidence of different responses of CO_2 efflux to environmental factors in different areas (Pangle and Seiler 2002; Zhang et al. 2006). Here, we discuss how environmental factors (temperature, moisture, and rainfall) influence CO_2 efflux in the arid region.

Our study showed a significantly positive correlation between CO_2 efflux and temperature, indicating that temperature was a limiting factor for soil respiration. This is in agreement with many studies that conducted in various ecosystems (Almagro et al. 2009; Godwin et al. 2017; Suh et al. 2009). For example, Yu et al. (2015) showed that soil temperature at 10 cm depth could explain 70.5% of variations for heterotrophic respiration in dryland agriculture system.

The dual effects of soil moisture on surface CO_2 efflux were also reported by Jassal et al. (2005) who found efflux in a forest soil showing an increase with increasing moisture to 12%, followed by a decrease with further increase in moisture. However, there was a positive correlation between CO_2 efflux and soil moisture reported in arid and semiarid region (Chang et al. 2009; Wu et al. 2010). The reason may be that soil water content can influence soil CO_2 flux in two directions. On the one hand,

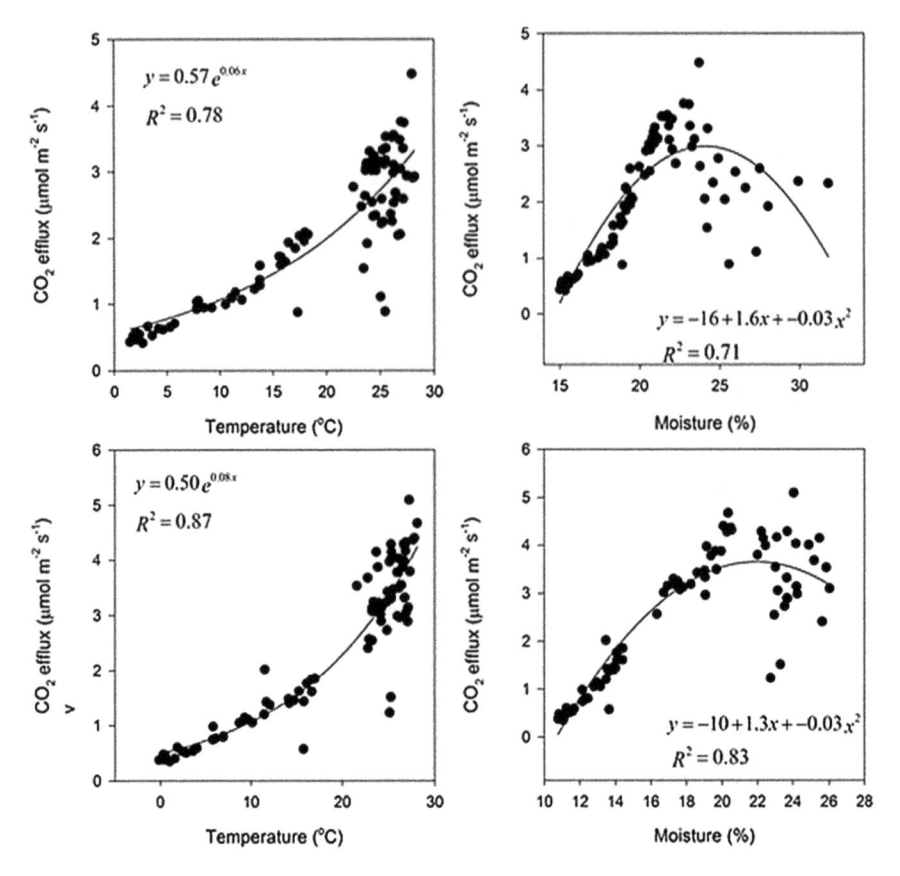

Fig. 6 Relationship between soil CO_2 efflux and temperature and moisture at 6 cm in the soil-only treatment (upper panel) at 8 cm in the soil with straw treatment (lower panel) using data from day 181 to day 324 (left panel redrawn from Wang et al. 2015a)

increasing soil moisture in dry soil can promote the decomposition of organic matter, especially in long dry season, as shown at a saline desert in western China (Ma et al. 2012) and in a semiarid grassland (Lopez-Ballesteros et al. 2016). As a result, there should be more CO_2 in soil profile and subsequent an increase in CO_2 efflux. On the other hand, more water in soil can cause more CO_2 to be dissolved (thus less CO_2 in soil profile) and also reduce air-filled porosity (Ball et al. 1999; Chayawat et al. 2012), leading to a decrease in CO_2 diffusivity and a subsequent decrease in CO_2 efflux.

The influence of rainfall on soil surface CO_2 efflux may be complex because various processes can be affected over different time scales. The two opposite effects linked with an increase in soil moisture (see discussion above) may last (from days to weeks) until soil water content becoming normal (i.e., close to average or that before rainfall or being stable). On the other hand, there may be other processes with short-term effects (from minutes to hours). For example, rainwater can fill the pore in soil profile and squeeze air (including CO_2 gas) out, which leads to a temporary increase in CO_2 efflux (Jassal et al. 2005); rainfall may cause a pause for CO_2 efflux or influx, which has been observed in sandy soil due to a gradient of gas pressure between soil–atmosphere interface (Fa et al. 2015).

4.2 Comparison of Different Diffusion Models

Our study demonstrates that the Marshall (1959) model may do a better job in estimating CO_2 efflux of the arid soil than the Penman (1940) and Moldrup et al. (2013) models although the Marshall (1959) model underestimates the seasonal amplitude. There have been reports that show differences between measured surface CO_2 efflux and gradient method using different diffusion models. For example, Pingintha et al. (2010) found that the Penman (1940) and Marshall (1959) models significantly overestimated CO_2 efflux in a non-irrigated peanut field. A study by Fan and Jones (2014) showed that the Moldrup et al. (2013) model overestimated CO_2 diffusion in a pasture plot with calcareous soil in a semiarid region, Utah, USA. However, Schwen et al. (2015) found that the Moldrup et al. (2013) model with modified parameters could reproduce the observed CO_2 diffusion in a maize field with calcareous soil under a sub-humid climate in lower Austria.

There have been studies using other models to predict CO_2 efflux. For example, Risk et al. (2015) applied the complex model by Mccarthy and Johnson (1995) that includes CO_2 diffusion in both free air and water and found that the model overpredicted (underpredicted) CO_2 diffusivity under dry (wet) condition in representative soils with a range of textural classes. Interestingly, Fan and Jones (2014) tested a simple power model by Buckingham (1904) that only considers air-filled porosity, but found that the model "was sufficient to provide estimates that agreed well with experimental measurements." In general, empirical models tend to have bigger error in relatively dry or wet situation, implying that those models may not be able to predict CO_2 efflux during and following rainfall. Thus, there is no single diffusion model that can perform well for all soil conditions.

5 Conclusions

Soil CO_2 concentration and efflux had a clear seasonal variation (with two peaks in summer and continuously decreasing in autumn). Surface CO_2 efflux had significantly positive correlation with soil temperature throughout the summer and autumn; soil CO_2 efflux linearly increased with increase of soil moisture when soil was not too wet (moisture less than ~22%).

The Marshall (1959) model showed a better performance than the Penman (1940) and Moldrup et al. (2013) models in the arid soil of Yanqi Basin. However, the

Marshall (1959) model underestimates the seasonal amplitude. Studies are needed to investigate the relationship between CO_2 flux and soil environmental variables under various conditions.

References

Almagro M, Lopez J, Querejeta JI, Martinez-Mena M (2009) Temperature dependence of soil CO_2 efflux is strongly modulated by seasonal patterns of moisture availability in a Mediterranean ecosystem. Soil Biol Biochem 41:594–605. https://doi.org/10.1016/j.soilbio.2008.12.021

Ball BC, Scott A, Parker JP (1999) Field N_2O, CO_2 and CH_4 fluxes in relation to tillage, compaction and soil quality in Scotland. Soil Till Res 53:29–39

Buckingham E (1904) Contributions to our knowledge of the aeration of soils. Bur Soil Bull, vol 25. U.S. Government Printing Office, Washington, DC

Chang ZQ, Feng Q, Su YH, Si JH, Xi HY, Cao SK, Guo R (2009) Influence of environmental, root, and stand parameters on soil surface CO_2 efflux in a Populus euphratica of desert forest in extreme arid region. Sci Cold Arid Reg 1:348–355

Chayawat C, Senthong C, Leclerc MY, Zhang GS, Beasley JP (2012) Seasonal and post-rainfall dynamics of soil CO_2 efflux in wheat and peanut fields. Chiang Mai J Sci 39:410–428

Fa KY, Liu JB, Zhang YQ, Wu B, Qin SG, Feng W, Lai ZR (2015) CO_2 absorption of sandy soil induced by rainfall pulses in a desert ecosystem. Hydrol Process 29:2043–2051. https://doi.org/10.1002/hyp.10350

Fan J, Jones SB (2014) Soil surface wetting effects on gradient-based estimates of soil carbon dioxide efflux. Vadose Zone J 13. https://doi.org/10.2136/vzj2013.07.0124

Fierer N, Chadwick OA, Trumbore SE (2005) Production of CO_2 in soil profiles of a California annual grassland. Ecosystems 8:412–429. https://doi.org/10.1007/s10021-003-0151-y

Godwin D, Kobziar L, Robertson K (2017) Effects of fire frequency and soil temperature on soil CO_2 efflux rates in Old-Field Pine-Grassland Forests. Forests 8:274

Jassal R, Black A, Novak M, Morgenstern K, Nesic Z, Gaumont-Guay D (2005) Relationship between soil CO_2 concentrations and forest-floor CO_2 effluxes. Agric For Meteorol 130:176–192. https://doi.org/10.1016/j.agrformet.2005.03.005

Jia X, Zha TS, Wu B, Zhang YQ, Gong JN, Qin SG, Chen GP, Qian D, Kellomaki S, Peltola H (2014) Abiotic CO_2 uptake from the atmosphere by semiarid desert soil and its partitioning into soil phases. Biogeosciences 11:4679–4693. https://doi.org/10.5194/bg-11-4679-2014

Jones HG (1992) Plants and microclimate: a quantitative approach to environmental plant physiology. Cambridge University Press, New York

Jone ED, Schapper HJ (1972) Calculation of soil respiration and activity from CO_2 profiles in soil. Soil Sci 113:328–333

Kammann C, Grunhage L, Jager HJ (2001) A new sampling technique to monitor concentrations of CH(4), N(2)O and CO(2) in air at well-defined depths in soils with varied water potential. Eur J Soil Sci 52:297–303. https://doi.org/10.1046/j.1365-2389.2001.00380.x

Li Y, Wang YG, Houghton RA, Tang LS (2015) Hidden carbon sink beneath desert. Geophys Res Lett 42:5880–5887. https://doi.org/10.1002/2015GL064222

Liu JB, Fa KY, Zhang YQ, Wu B, Qin SG, Jia X (2015) Abiotic CO_2 uptake from the atmosphere by semiarid desert soil and its partitioning into soil phases. Geophys Res Lett 42:5779–5785. https://doi.org/10.1002/2015GL064689

Liu Z, Zhang YQ, Fa KY, Qin SG, She WW (2017) Rainfall pulses modify soil carbon emission in a semiarid desert. CATENA 155:147–155. https://doi.org/10.1016/j.catena.2017.03.011

Lopez-Ballesteros A, Serrano-Ortiz P, Sanchez-Canete EP, Oyonarte C, Kowalski AS, Perez-Priego O, Domingo F (2016) Enhancement of the net CO_2 release of a semiarid grassland in SE Spain by rain pulses. J Geophys Res-Biogeo 121:52–66. https://doi.org/10.1002/2015JG003091

Ma J, Zheng XJ, Li Y (2012) The response of CO_2 flux to rain pulses at a saline desert. Hydrol Process 26:4029–4037. https://doi.org/10.1002/hyp.9204

Maier M, Schack-Kirchner H, Hildebrand EE, Holst J (2010) Pore-space CO_2 dynamics in a deep, well-aerated soil. Eur J Soil Sci 61:877–887. https://doi.org/10.1111/j.1365-2389.2010.01287.x

Marshall TJ (1959) The diffusion of gases through porous media. J Soil Sci 10:79–82

Mccarthy KA, Johnson RL (1995) Measurement of trichloroethylene diffusion as a function of moisture-content in sections of gravity-drained soil columns. J Environ Qual 24:49–55

Moldrup P, Deepagoda TKKC, Hamamoto S, Komatsu T, Kawamoto K, Rolston DE, de Jonge LW (2013) Structure-dependent water-induced linear reduction model for predicting gas diffusivity and tortuosity in repacked and intact soil. Vadose Zone J 12. https://doi.org/10.2136/vzj2013.01.0026

Pangle RE, Seiler J (2002) Influence of seedling roots, environmental factors and soil characteristics on soil CO_2 efflux rates in a 2-year-old loblolly pine (Pinus taeda L.) plantation in the Virginia Piedmont. Environ Pollut 116:S85–S96. https://doi.org/10.1016/S0269-7491(01)00261-5

Pattey E, Strachan IB, Desjardins RL, Massheder J (2002) Measuring nighttime CO_2 flux over terrestrial ecosystems using eddy covariance and nocturnal boundary layer methods. Agric For Meteorol 113:145–158. https://doi.org/10.1016/S0168-1923(02)00106-5

Penman HL (1940) Gas and vapour movements in the soil I. The diffusion of vapours through porous solids. J Agric Sci 30:437–462

Pingintha N, Leclerc MY, Beasley JP, Zhang GS, Senthong C (2010) Assessment of the soil CO_2 gradient method for soil CO_2 efflux measurements: comparison of six models in the calculation of the relative gas diffusion coefficient. Tellus B 62:47–58. https://doi.org/10.1111/j.1600-0889.2009.00445.x

Risk D, Kellman L, Beltrami H (2015) A new method for in situ soil gas diffusivity measurement and applications in the monitoring of subsurface CO_2 production. J Geophys Res Biogeosciences 113:96

Sanderman J, Amundson R (2010) Soil carbon dioxide production and climatic sensitivity in contrasting California ecosystems. Soil Sci Soc Am J 74:1356–1366. https://doi.org/10.2136/sssaj2009.0290

Schwen A, Jeider E, Bottcher J (2015) Spatial and temporal variability of soil gas diffusivity, its scaling and relevance for soil respiration under different tillage. Geoderma 259:323–336. https://doi.org/10.1016/j.geoderma.2015.04.020

Suh S, Lee E, Lee J (2009) Temperature and moisture sensitivities of CO_2 efflux from lowland and alpine meadow soils. J Plant Ecol 2:225–231. https://doi.org/10.1093/jpe/rtp021

Tang JW, Baldocchi DD, Qi Y, Xu LK (2003) Assessing soil CO_2 efflux using continuous measurements of CO_2 profiles in soils with small solid-state sensors. Agric For Meteorol 118:207–220. https://doi.org/10.1016/S0168-1923(03)00112-6

Tang JW, Misson L, Gershenson A, Cheng WX, Goldstein AH (2005) Continuous measurements of soil respiration with and without roots in a ponderosa pine plantation in the Sierra Nevada Mountains. Agric For Meteorol 132:212–227. https://doi.org/10.1016/j.agrformet.2005.07.011

Wang TT, Wang XJ, Zhao CY, Wang JP, Yu ZT, Zhang J (2015a) Characteristics of soil respiration in bare soil of farmland in the South of Xinjiang. Arid Zone Res 32:453–460

Wang JH, Wang XJ, Xu MG, Feng G, Zhang WJ, Lu CA (2015b) Crop yield and soil organic matter after long-term straw return to soil in China. Nutr Cycl Agroecosys 102:371–381. https://doi.org/10.1007/s10705-015-9710-9

Wang M, Guan DX, Han SJ, Wu JL (2010) Comparison of eddy covariance and chamber-based methods for measuring CO_2 flux in a temperate mixed forest. Tree Physiol 30:149–163. https://doi.org/10.1093/treephys/tpp098

Wiaux F, Vanclooster M, Van Oost K (2015) Vertical partitioning and controlling factors of gradient-based soil carbon dioxide fluxes in two contrasted soil profiles along a loamy hillslope. Biogeosciences 12:4637–4649. https://doi.org/10.5194/bg-12-4637-2015

Wu X, Yao Z, Bruggemann N, Shen ZY, Wolf B, Dannenmann M, Zheng X, Butterbach-Bahl K (2010) Effects of soil moisture and temperature on CO_2 and CH4 soil atmosphere exchange of various land use/cover types in a semi-arid grassland in Inner Mongolia, China. Soil Biol Biochem 42:773–787. https://doi.org/10.1016/j.soilbio.2010.01.013

Xie JX, Li Y, Zhai CX, Li CH, Lan ZD (2009) CO_2 absorption by alkaline soils and its implication to the global carbon cycle. Environ Geol 56:953–961. https://doi.org/10.1007/s00254-008-1197-0

Yan MF, Zhou GS, Zhang XS (2014) Effects of irrigation on the soil CO_2 efflux from different poplar clone plantations in arid northwest China. Plant Soil 375:89–97. https://doi.org/10.1007/s11104-013-1944-1

Yu Y, Zhao C, Zhao Z, Yu B, Zhou T (2015) Soil respiration and the contribution of root respiration of cotton (Gossypium hirsutum L.) in arid region. Acta Ecol Sin 35:17–21

Zhang DQ, Sun XM, Zhou GY, Yan JH, Wang YS, Liu SZ, Zhou CY, Liu JX, Tang XL, Li J, Zhang QM (2006) Seasonal dynamics of soil CO_2 effluxes with responses to environmental factors in lower subtropical forests of China. Sci China Ser D 49:139–149. https://doi.org/10.1007/s11430-006-8139-z

Zhang LH, Chen YN, Zhao RF, Li WH (2010) Significance of temperature and soil water content on soil respiration in three desert ecosystems in Northwest China. J Arid Environ 74:1200–1211. https://doi.org/10.1016/j.jaridenv.2010.05.031

Land Use Impacts on Soil Organic and Inorganic Carbon and Their Isotopes in the Yanqi Basin

Jiaping Wang, Xiujun Wang and Juan Zhang

Abstract Assessments of both soil organic carbon (SOC) and inorganic carbon (SIC) under different land use types are lacking in arid regions. To advance the understanding of soil carbon dynamics and impacts of land use changes, a study was conducted in the central Xinjiang, the Yanqi Basin that had significant land use changes since the 1950s. Soil samples were collected at representative sites under various types of land use and vegetation, and SOC and SIC and their isotopes were measured over the 0–100 cm. This study revealed that both SOC and SIC stocks were the lowest in the desert land, but the highest in the agricultural land. Conversion of native land to cropland has caused a significant increase of SOC in the topsoil, and SIC in the subsoil. Total soil carbon stocks in the 0–100 cm are 11.6 ± 4.8, 44.7 ± 10.4, and 51.2 ± 5.6 kg C m^{-2} in the desert land, shrub land, and agricultural land, respectively. On average, soil inorganic carbon counts more than 80% of the total carbon stock in the soil profiles. δ^{13}C of SOC shows no significant differences between land use types, but δ^{13}C of SIC is much different among land use types, following an order: desert land (−0.6‰) > shrub land (−2.2‰) > agricultural land (−3.4‰). Our finding of SIC increase with depletion of ^{13}C in the agricultural land indicates that there has been enhanced accumulation of pedogenic carbonate as a result of cropping.

J. Wang (✉)
College of Agriculture, Shihezi University, Shihezi City, Xinjiang 832000, China
e-mail: 2006wjp@163.com

X. Wang
College of Global Change and Earth System Science, Beijing Normal University, Beijing 100875, China

J. Zhang
School of Resources and Environment, Northeast Agricultural University, Harbin 150030, China

1 Introduction

There have been many studies on the terrestrial carbon cycle over the last century because of the great concern on the continuous increase of atmospheric carbon dioxide (CO_2), which may be the main reason for the global warming. Soil organic carbon (SOC), the largest carbon pool in the terrestrial ecosystem, has attracted great attention (Sanderman et al. 2008; Vargas et al. 2011). In contrast, soil inorganic carbon or carbonate (SIC), mainly found in the arid and semi-arid lands, may also play an important part in the global carbon cycle and climate change (Entry et al. 2004; Lal 2004; Schlesinger 1999). Apparently, studying the dynamics of both SOC and SIC is crucial to the understanding of the carbon cycle over the regional and global scales.

There have been significant land use changes (i.e., conversion of native lands to agricultural lands with fertilization and irrigation) in the arid and semi-arid regions of China over the past decades, which may have implications for the carbon cycling. A few studies have indicated that there is an increase in SIC in irrigated agricultural lands in arid regions, relative to non-irrigation lands (Wu et al. 2008, 2009; Zhang et al. 2010). While the topic of soil carbon storage has been a focus of the carbon cycle research (Gang et al. 2012; Su et al. 2010; Wu et al. 2008), there are limited studies of both SOC and SIC profiles (Wang et al. 2010; Yang et al. 2007), and little is done to assess impacts of land use change on soil carbon dynamics in northwest China.

Because of ^{13}C discrimination during photosynthesis, different plant has its own isotopic signature. Thus, vegetation type can have influences not only on soil carbon stocks but also on their stable isotopic compositions. It is well documented that C3 and C4 plants have much different isotopic signatures, with a mean $\delta^{13}C$ value of −27 and −12‰, respectively (Cerling 1984). Thus, $\delta^{13}C$ in SOC gives us a way to trace vegetation change from C3 to C4, or vice versa (Boutton et al. 1998; Krull and Bray 2005). On the other hand, $\delta^{13}C$ in SIC may be used to differentiate carbonate origination, i.e., lithogenic carbonate (LIC) and pedogenic carbonate (PIC) (Liu et al. 1996; Nordt et al. 1998).

There are a few methods for determining SOC and SIC. For example, dry combustion with automated analyzers and wet chemical oxidation are common methods for SOC measurements (Walkley and Black 1934). While the automated technique is quick and accurate, the cost is usually high. Commonly, SIC is often measured by determining the CO_2 production following the addition of HCl acid (Dreimanis 1962; Heiri et al. 2001; Presley 1975; Sherrod et al. 2002). Another simple and cheap method is the loss-on-ignition (LOI) that is to heat soils at high temperature to cause combustion of soil organic material (SOM) or carbonate and measure the weight losses (Wang et al. 1996, 2011b). Compared to the automated techniques and chemical methods, LOI techniques are less expensive and labor intensive (Wang et al. 2012).

The Yanqi Basin has had land use changes over the past decades. Majority of the shrub land in the Yanqi Basin was changed to agricultural land that has been used for farming. A set of studies were carried out to understand the impacts of land use changes on the dynamics of soil carbon, which include the assessments of common methods for SOC and SIC measurements, and evaluations of the profile distributions of SOC and SIC and their isotopic compositions under various land uses.

2 Materials and Methods

2.1 Site Description and Soil Sampling

The Yanqi Basin is located in the northeast brim of the Taklamakan Desert, China. The main soil types are brown desert soil and gray-brown desert soil, which were developed from limestone parent material and classified as a Haplic Calcisol (FAO-UNESCO 1988). There are various land uses, with approximately 70% as agricultural land and ~30% as shrub land and desert. Due to the extremely low precipitation, agriculture land uses water from the Kaidu River; rainfall and underground waters are the water sources for the desert and shrub lands.

We randomly selected 21 sites in the Yanqi Basin during August and November, 2010: 3 sites from desert, 9 sites from shrub land, and 9 sites from agricultural land (Fig. 1). We collected soils over five layers at most profiles in the desert and shrub lands, and six layers in the agricultural land. Initially, soil was air-dried, well mixed, and then sieved to pass a 2-mm screen. Bulk density was determined for all sites and soil texture for representative sites. Soil texture was determined for surface soil samples (0–30 cm) using the Mastersizer 2000 particle size analyzers. Soil pH and electrical conductivity (EC) were determined using a soil:water (1:5) mixture. Representative sub-samples were ground to 0.25 mm for measuring SOC and SIC and their isotopes.

2.2 Basic Soil Properties

Table 1 shows the basic properties of the topsoil (0–30 cm). In general, clay content was low (3–8%) in all land uses, with the highest value in the agricultural soils. Sand content was significantly higher in the desert soils (68%) than in the shrub (25%) and agricultural lands (20%). Silt content was much lower in the desert soils (29%) than in the shrub and agricultural soils (69–72%). Bulk density was the highest in desert land (1.6 g cm^{-3}) and lowest in the agricultural land (1.3 g cm^{-3}). Soil pH was similar between the desert (8.3), shrub (8.4) and agricultural lands (8.2). However, EC was much higher in the shrub land (9.1 ms cm^{-1}) than in the desert and agricultural lands (0.8–1.4 ms cm^{-1}).

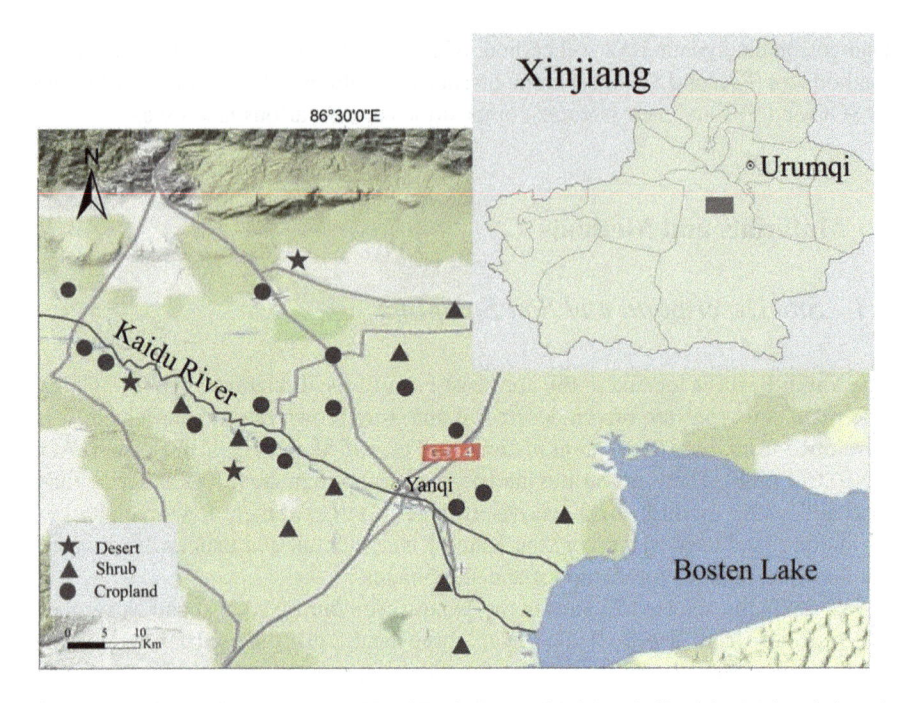

Fig. 1 Map of sampling layout in the Yanqi Basin [Desert land (stars), Shrub land (triangles), and Agricultural land (circles)] (redrawn from Wang et al. 2012)

Table 1 Means (standard deviations) of soil basic properties in desert land (DL), shrub land (SL), and agricultural land (AL) (redrawn from Wang et al. 2012)

	BD	pH	EC	Clay	Silt	Sand
	g cm^{-3}		ms cm^{-1}	%		
DL	1.6 (0.2)	8.3 (0.2)	1.4 (2.0)	3.1 (0.5)	28.8 (12.1)	68.1 (12.2)
SL	1.4 (0.1)	8.4 (0.3)	9.1 (7.8)	5.5 (1.8)	69.4 (9.8)	25.1 (10.8)
AL	1.3 (0.1)	8.2 (0.2)	0.8 (0.6)	7.6 (1.6)	72.3 (4.4)	20.1 (5.8)

2.3 Soil Carbon Measurements

2.3.1 Traditional Methods

We used a modified Walkley–Black method to determine SOC, which was modified according to the traditional Walkley–Black method (i.e., Walkley and Black 1934). In brief, soil was treated with concentrated H_2SO_4 and 0.5 M $K_2Cr_2O_7$ at 150–160 °C, then followed by titrating with 0.25 M $FeSO_4$. The pressure calcimeter method was used for SIC measurement, which was similar to that of Sherrod et al. (2002).

2.3.2 LOI Methods

Based on the methods by Wang et al. (1996) and Wang et al. (2011b), we first heated soil at 105 °C for 24 h to remove soil moisture. We then follow a two-step procedure for both SOC and SIC estimates by combusting soil in a muffle furnace (S1849, KOYO LINDBEERE LTD): At 375 °C for 17 h, then at 800 °C for 12 h. Soil organic matter is calculated as the weight difference between 105 and 375 °C, and SIC as the weight loss during the combustion under 800 °C.

2.3.3 Elemental Analyzer

We also used a CNHS-O analyzer (EuroEA3000) to determine SOC and SIC contents. For SOC estimate, soil is pretreated with 10 drips of H_3PO_4 for 12 h to remove carbonate. Then the sample is combusted and CO_2 production is determined by a thermal conductivity detector. Total soil carbon is measured, using the same procedure without pretreatment with H_3PO_4. We calculated the difference between total soil carbon and SOC as SIC content.

2.3.4 Carbon Isotope Measurement

Stable isotope, $\delta^{13}C$, in SOC was measured using a Finigan MAT Delta Plus XP at the State Key Laboratory of Lake Science and Environment (SKLLSE), Nanjing Institute of Geography and Limnology, Chinese Academy of Sciences (CAS), and $\delta^{13}C$ in SIC at the Nanjing Institute of Geology and Paleontology, CAS. For $\delta^{13}C$ in SOC, soil was pretreated with H_3PO_4 to remove carbonate. The pretreated soil was combusted at 1020 °C with constant helium flow carrying pure oxygen to ensure complete oxidation of organic materials. For $\delta^{13}C$ in SIC, CO_2 was collected during the reaction of soil with 100% H_3PO_4 and transported in a helium stream to the mass spectrometer. Isotopic data were reported in delta notation relative to the Vienna Pee Dee Belemnite (VPDB).

2.4 Evaluations of Soil Carbon Methods

2.4.1 Comparisons of SOC Methods

Figure 2 shows the comparison in SOC measurement between the modified Walkley–Black method and the automated CNS analysis. Clearly, there is a strong linear relationship between the two methods. The regression line is forced through zero since the intercept is not significantly different from zero. The slope value of 1.023 is significantly different from 1. When SOC content is too low (less the 5 g kg^{-1}), the modified Walkley–Black method can cause over-estimate for SOC. If the low-SOC

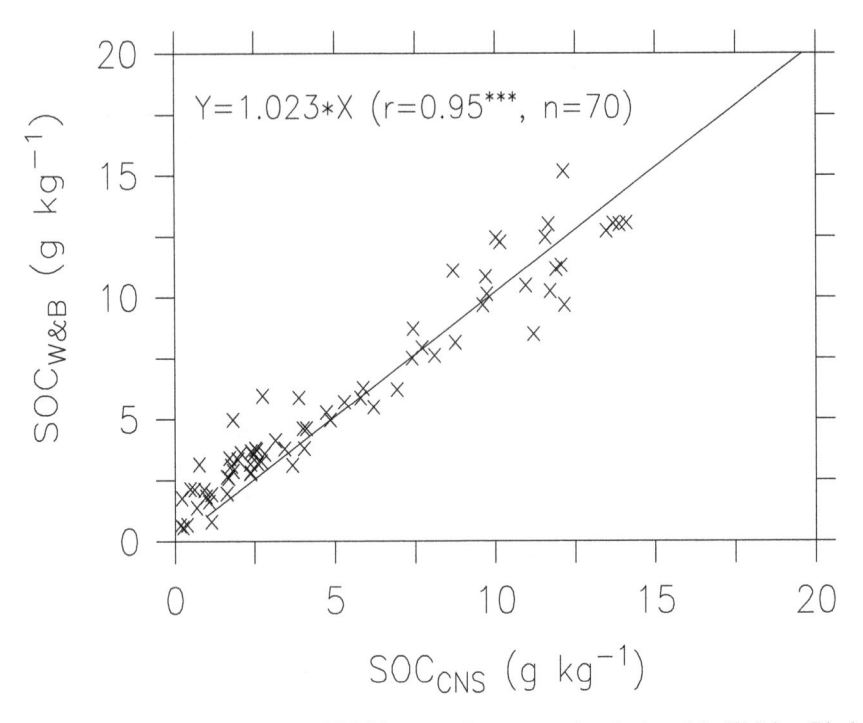

Fig. 2 Linear regression in measured SOC between the automated analysis and the Walkley–Black method (redrawn from Wang et al. 2012)

data are excluded, a new regression line gives a slope value of 0.997. These comparisons demonstrate that near 100% SOC can be oxidized in the soils in this region, which is much higher than previous published values for forest soils in Flanders (De Vos et al. 2007), but in line with those (~100%) in the Tasmanian soils (Wang et al. 1996). The high SOC recovery and large correlation coefficient (0.95) in this study indicate that the modified Walkley–Black method can provide good estimates of SOC, except for soils with very low levels of SOC.

Figure 3 shows the relationship between LOI-derived SOM and automated SOC by the CNS analyzer. There is a large intercept (4.189) for the regression line, suggesting some weight losses of non-SOM materials (e.g., structural water and carbonate) during ignition. The conversion factor (1.792) is little larger than the traditional value (1.724). Accordingly, SOC can be estimated by LOI at 375 °C for 17 h as follows:

$$SOC_{LOI} = (SOM_{LOI} - 4.189)/1.792 \tag{1}$$

As demonstrated in Table 2, the calculated SOC using the LOI method (i.e., above equation) has a range of 0.1–13.3 g kg^{-1}, mean value of 6.2 g kg^{-1}, and standard

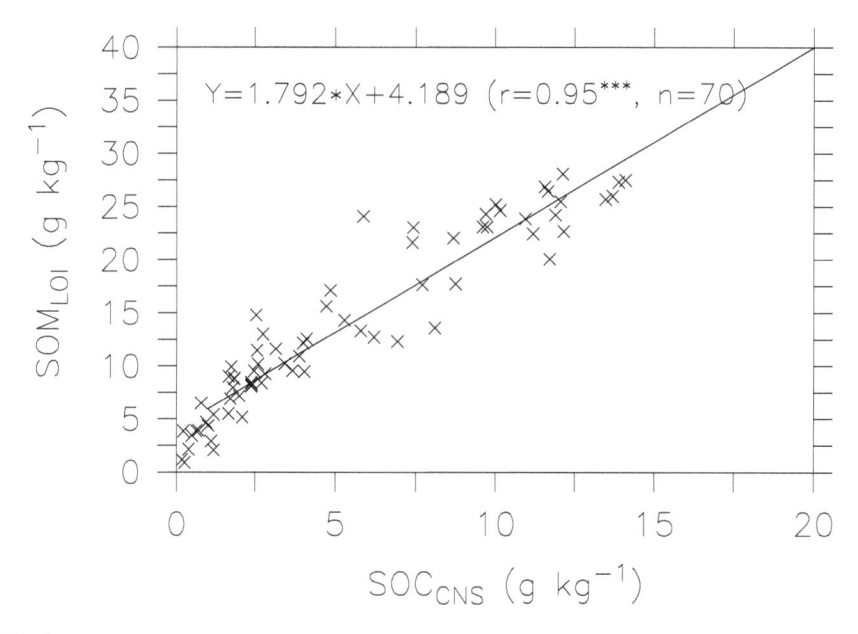

Fig. 3 Linear regression between measured SOC by the CNS Analyzer and LOI-derived SOM (redrawn from Wang et al. 2012)

Table 2 Analyses of soil organic carbon (SOC) (g kg^{-1}) and inorganic carbon (SIC) (g kg^{-1}) determined by three different methods (redrawn from Wang et al. 2012)

	SOC$_{CNS}$	SOC$_{W\&B}$	SOC$_{LOI}$	SIC$_{CNS}$	SIC$_{PC}$	SIC$_{LOI}$
Maximum	14.11	15.17	13.32	52.97	51.47	54.93
Minimum	0.78	1.84	0.10	8.53	6.34	9.26
Mean	6.00	6.57	6.20	24.65	23.55	28.28
Standard deviation	4.19	3.79	4.18	10.39	10.73	11.07

deviation of 4.2 g kg^{-1} for the soils tested, which are similar to those (0.8–14.1, 6.0, and 4.2 g kg^{-1}, respectively) determined by the elemental CNS analyzer.

2.4.2 Comparisons of SIC Methods

SIC determined by the pressure calcimeter method shows significantly linear relationship ($r = 0.99$, $p < 0.001$) with the automated CNS analysis (Fig. 4) although the former yields slightly lower values than the latter (Table 2). Since the intercept is not significantly different from zero, the regression line is forced through the origin. The close-to-1 slope value and large correlation coefficient give us confidence that the pressure calcimeter method is comparable with the automated CNS technique for the tested calcareous soils.

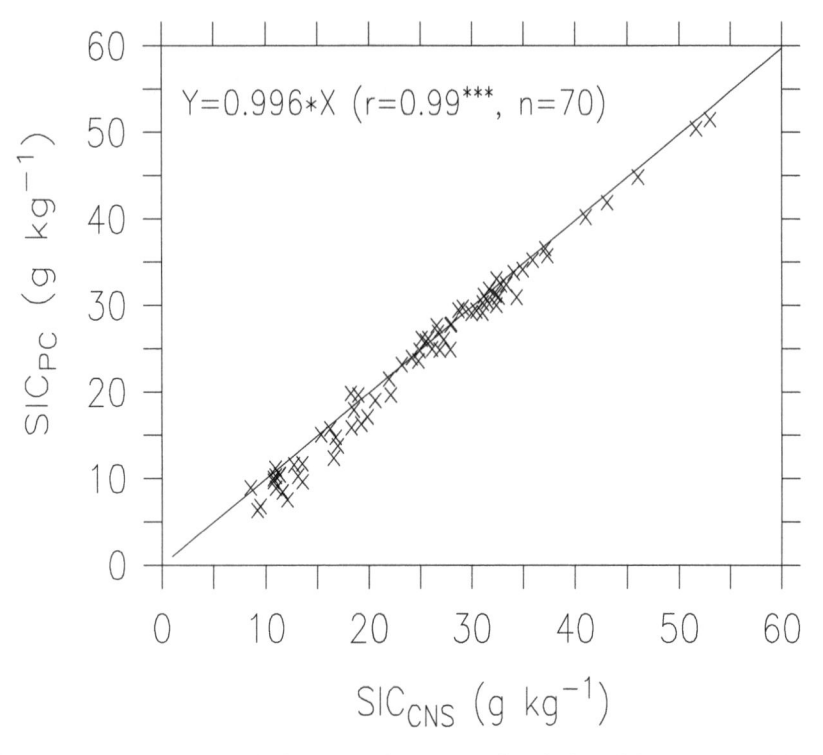

Fig. 4 Linear relationship for SIC between the automated analysis and the pressure calcimeter method (redrawn from Wang et al. 2012)

As shown by Fig. 5, the LOI method produces a very similar range of SIC to the automated CNS analysis, and there is a significant linear relationship between the two methods:

$$SIC_{LOI} = 1.043\,SIC_{CNS} + 2.582 \qquad (2)$$

The intercept value of 2.582 is significantly bigger than 0, indicating that there may be weight losses of non-SIC materials (e.g., dehydration from other compounds) during ignition (Santisteban et al. 2004). However, the slope value is close to 1 and correlation is significant ($r = 0.98$, $P < 0.001$), which suggest that SIC content can be accurately measured by the LOI technique.

Figure 6 shows that there is also a strong linear relationship between the pressure calcimeter method and the LOI technique for SIC:

$$SIC_{LOI} = 1.021\,SIC_{PC} + 4.241, \qquad (3)$$

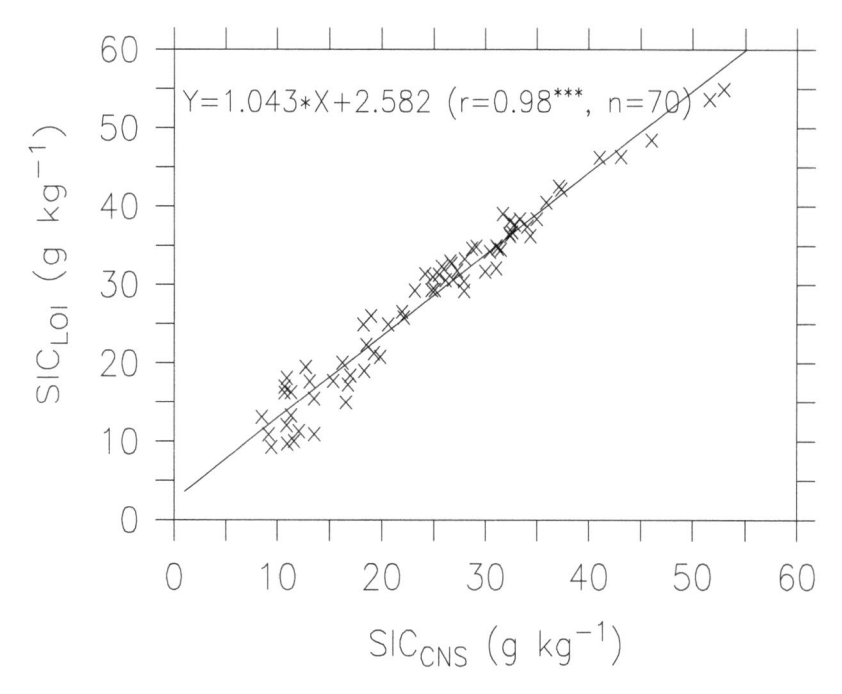

Fig. 5 Linear relationship between the automated analysis and the LOI method for SIC (redrawn from Wang et al. 2012)

which is similar to the regression between the LOI-SIC and the CNS-SIC (see Eq. 2). The much higher intercept (4.241) in Eq. (3) causes higher estimates for SIC by the LOI method, compared to those by the CNS technique (Table 1).

3 Results and Discussion

3.1 Soil Carbon and $\delta^{13}C$ Under Various Land Uses

3.1.1 Vertical Distributions of SOC and Its $\delta^{13}C$

We first assessed the vertical distributions of SOC and its $\delta^{13}C$ value at the desert sites (Fig. 7). As was expected, all the soil profiles show low-SOC content, i.e., less than 4 g kg^{-1} except at site D2(0–5 cm) that reveals a rapid decrease of SOC from ~6 g kg^{-1} in the topsoil to almost zero in the subsoil. Overall, $\delta^{13}C$ in SOC is more negative in the topsoil (close to −22‰) than in the subsoils (close to −19‰).

Soil organic carbon varies widely in the shrub land in terms of magnitude and vertical distribution (Fig. 8). While more than 50% of the sites show small vertical variations of SOC, there is an obvious decrease (from ~20 g kg^{-1} in the topsoil to

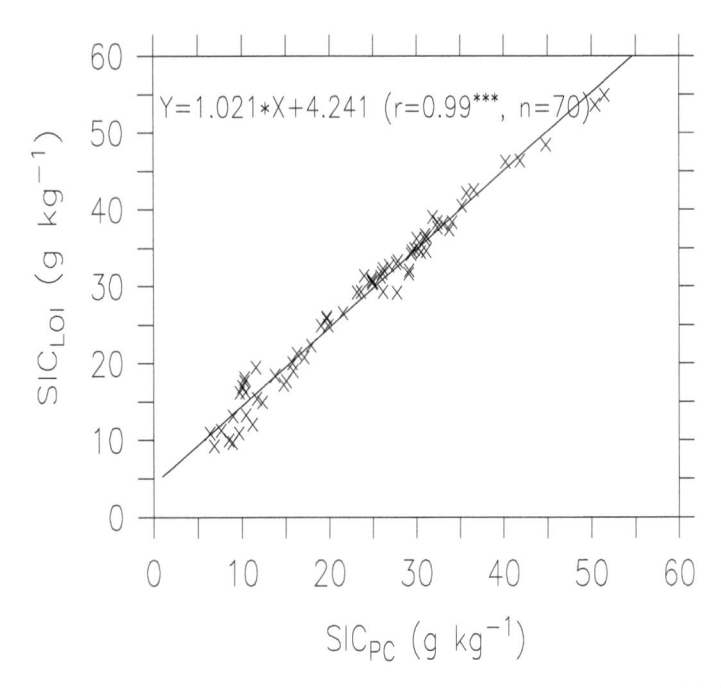

Fig. 6 Linear relationship between the pressure calcimeter and LOI methods for SIC (redrawn from Wang et al. 2012)

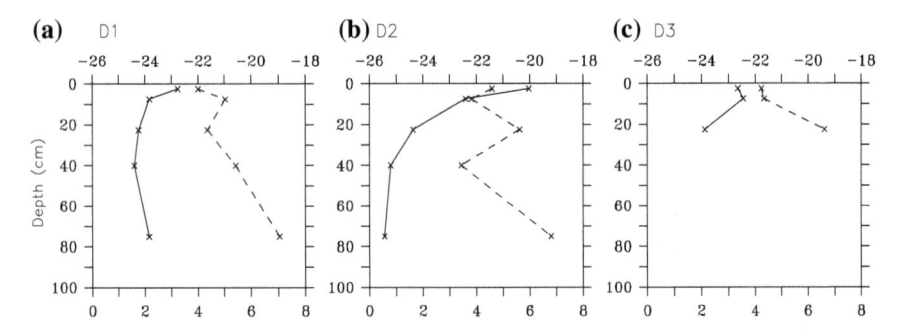

Fig. 7 Vertical distributions of SOC (solid line) and $\delta^{13}C$ (dashed line) on the desert land

<7 g kg^{-1} in the subsoil) in SOC in one-third of the profiles. It appeared that $\delta^{13}C$ in SOC is less negative in the subsoil at most sites, and there is an overall negative correlation between $\delta^{13}C$ and SOC content. For example, the vertical distribution of $\delta^{13}C$ is opposite to that of SOC at the sites S3, S4, and S5.

Figure 9 presents vertical profiles of SOC and $\delta^{13}C$ in the agricultural land. In general, SOC content in the topsoil is greater than 10 g kg^{-1} in the agricultural land, which is much larger than those in the desert and shrub lands. Generally, SOC vertical distribution on the agricultural land shows a sharp decline from 10 to 15 g kg^{-1} (above

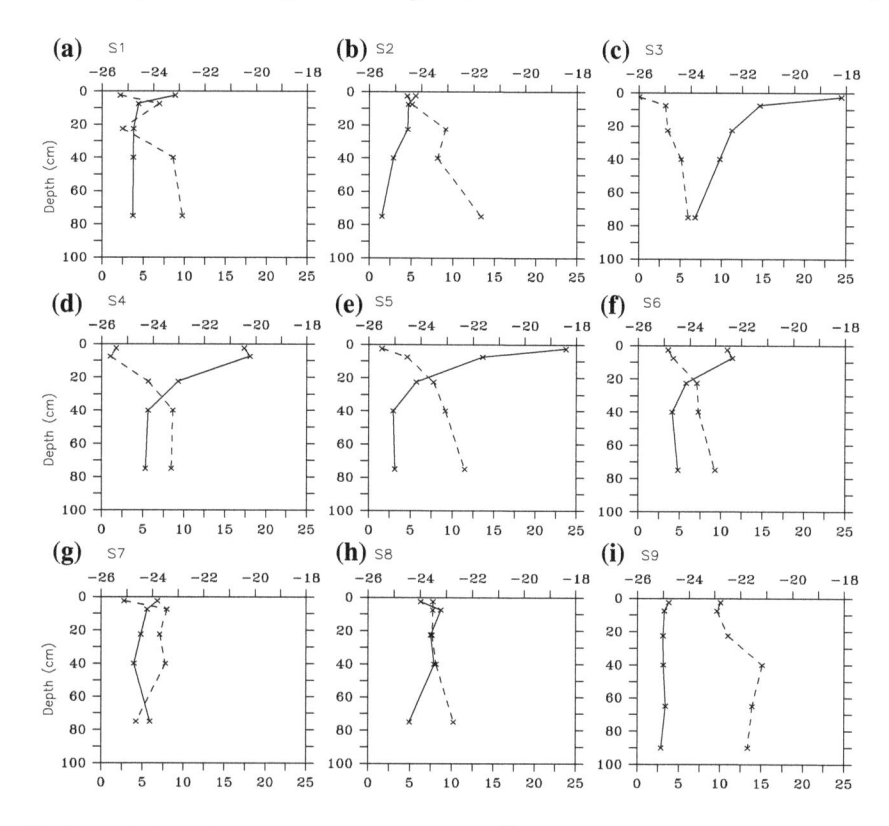

Fig. 8 Vertical distributions of SOC (solid line) and $\delta^{13}C$ (dashed line) on the shrub land

20 cm) to <4 g kg^{-1} (below 70 cm). The $\delta^{13}C$ value varies narrowly in the topsoil, from −25 to −23‰, but widely in the subsoil (from −24 to −20‰) in the agricultural land. Similar to the shrub land, there is enriched ^{13}C in the subsoil and a negative correlation between $\delta^{13}C$ and SOC at almost agricultural sites.

It is well known that C3 and C4 plants have much different isotopic values, with a mean $\delta^{13}C$ value of −27 and −12‰, respectively (Cerling 1984). Therefore, $\delta^{13}C$ in SOC provides us a way to trace land use history such as vegetation change between C3 and C4 plants (Boutton et al. 1998; McPherson et al. 1993). The value of $\delta^{13}C$ in SOC ranges from −20 to −27.5‰ in the Yanqi Basin, indicating that the dominant vegetation has been C3 type of plants. The $\delta^{13}C$ value becomes less negative with depth, implying two kinds of possible mechanisms. Firstly, the depletion of ^{13}C in the surface soil that contains newer SOM indicates an expansion of C3 vegetation in the recent years. Secondly, the enrichment of ^{13}C in the deep soil that has relatively older SOM may be caused by isotopic discrimination with SOC decomposition(Fernandez and Cadisch 2003; McPherson et al. 1993).

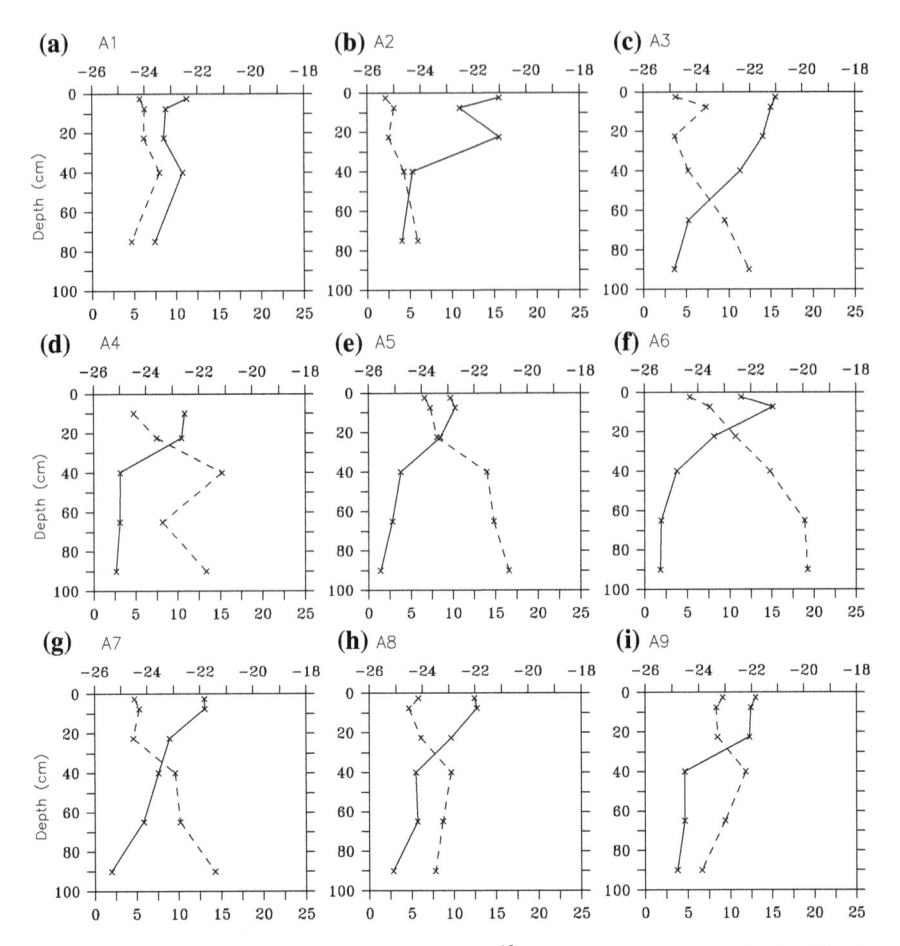

Fig. 9 Vertical distributions of SOC (solid line) and $\delta^{13}C$ (dashed line) on the agricultural land

3.1.2 Soil Inorganic Carbon and Its $\delta^{13}C$

Soil inorganic carbon and its $\delta^{13}C$ value show almost uniform distribution in soil profiles in the desert land (Fig. 10), which ranges from 8 to 14 g C kg^{-1} and −2 to 0‰, respectively. While SIC shows similar values(~10 g C kg^{-1}) in both D1 and D2 sites, $\delta^{13}C$ value in SIC is more negative at the site D1 (close to −2‰) than at D2 (close to 0).

As shown in Fig. 11, SIC content is generally higher in the shrub land (>15 g C kg^{-1}), except at S9 that reveals the lowest values for both SOC (<4 g C kg^{-1}) and SIC (~10 g C kg^{-1}). There is an increasing trend in SIC content with depth in more than half of the profiles, especially at the sites S3, S4, and S5, which have relatively higher SOC (>19 g C kg^{-1}) in the surface soil (see Fig. 8). The $\delta^{13}C$ value in SIC varies between −1 and −5‰ in the shrub land, with weak vertical variation in most

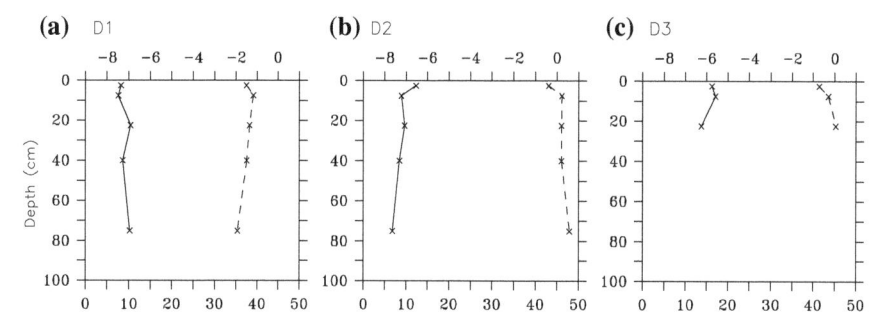

Land Use Impacts on Soil Organic and Inorganic ... 81

Fig. 10 Vertical distributions of SIC (solid line) and $\delta^{13}C$ (dashed line) on the desert land

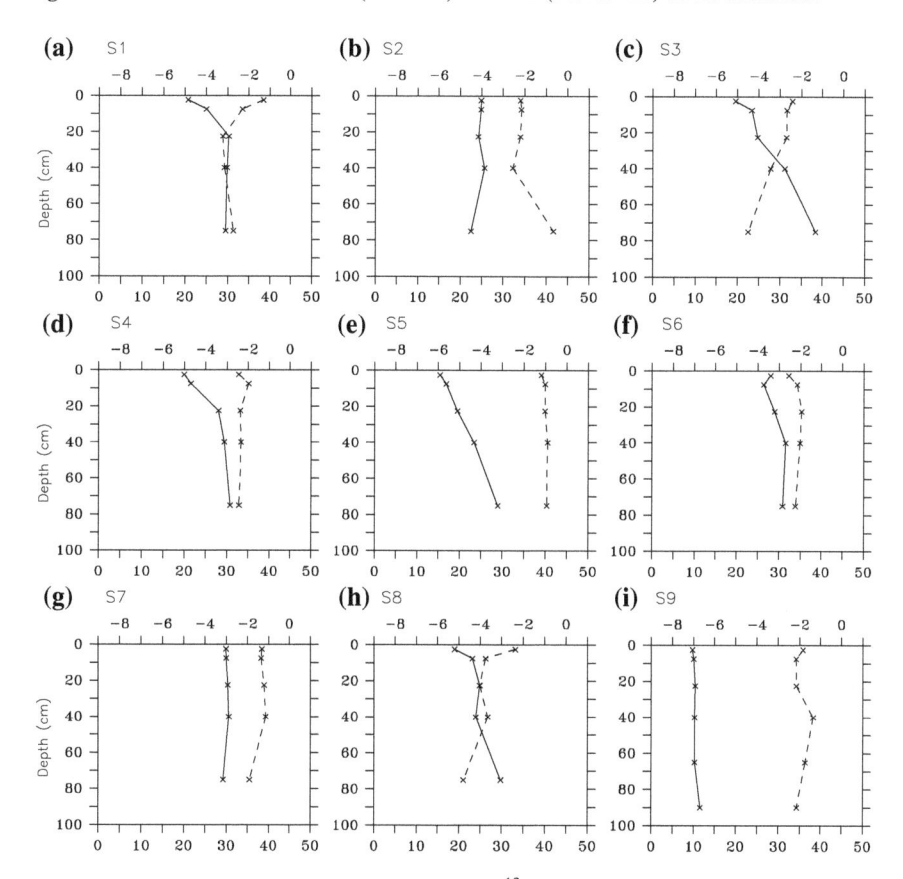

Fig. 11 Vertical distributions of SIC (solid line) and $\delta^{13}C$ (dashed line) on the shrub land

profiles. But there is a decline in $\delta^{13}C$ value of SIC with depth at the S1 and S3 where SIC shows an increasing trend with depth.

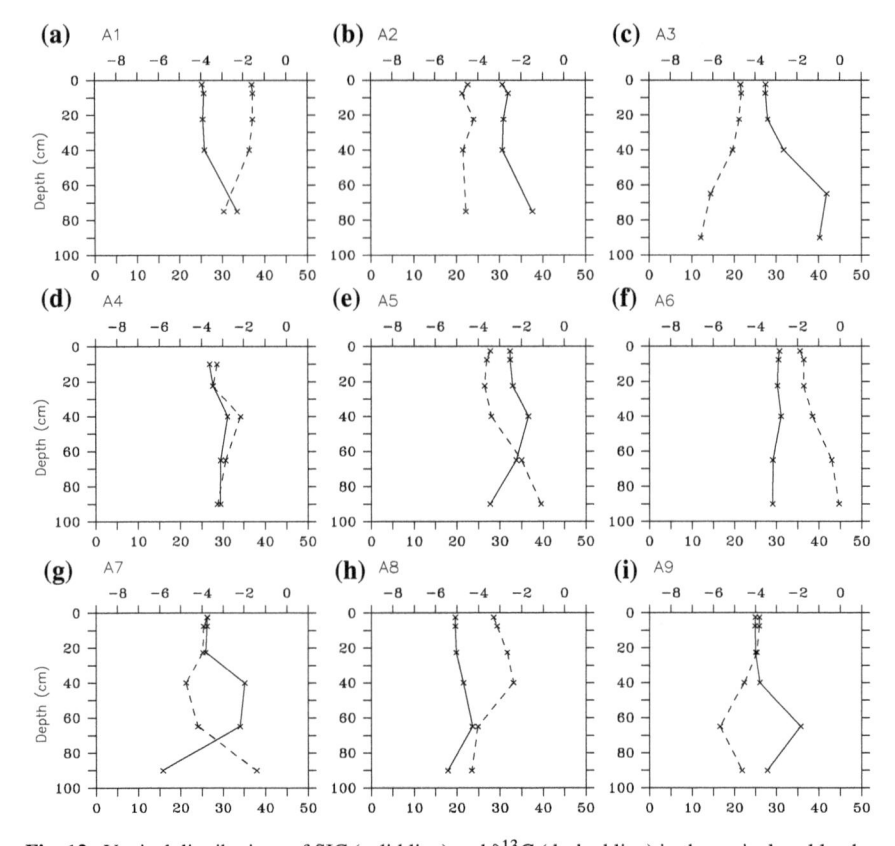

Fig. 12 Vertical distributions of SIC (solid line) and $\delta^{13}C$ (dashed line) in the agricultural land

Figure 12 presents SIC content and its $\delta^{13}C$ value in the agricultural land. Relative to the desert and shrub lands, SIC content is much higher, showing >25 g kg^{-1} in most profiles. Generally, SIC is uniformly distributed over the 0–30 cm layer. There are high SIC values ($\geq$35 g kg^{-1}) in the subsurface at some sites, i.e., sites A2, A3, A5, and A7. Similar to SIC content, the $\delta^{13}C$ value in SIC shows little vertical variation above 30 cm. There is a big range in the $\delta^{13}C$ value, from −5 to −1.5‰ over the 0–30 cm, and −6.5‰ to ~0 in the lower soil. Overall, there is a negative correlation between $\delta^{13}C$ and SIC in the agricultural land, i.e., the higher SIC content, the lower $\delta^{13}C$ value is.

3.1.3 Comparisons of Soil Carbon and the Isotopes Between Land Uses

Table 3 shows the SOC, SIC, and their stable isotopes for all individual layers. Statistical analyses demonstrate that both SOC and SIC are significantly higher in the shrub land and agricultural land than in the desert land. In spite of the lacking of

Table 3 Average (standard variations) of soil carbon (g C kg^{-1}) and stable isotopes (‰) for the desert land (DL), shrub land (SL), and agricultural land (AL)

Depth (cm)	Land use	SOC	SIC	$\delta^{13}C_{SOC}$	$\delta^{13}C_{SIC}$
0–5	DL	4.2 (1.5) a	12.3 (4.0) a	−21.72 (0.3) a	−0.87 (0.58) a
	SL	11.9 (7.9) b	20.8 (6.2) b	−24.78 (1.0) b	−1.95 (0.6) a
	AL	12.8 (2.0) b	27.2 (4.1) c	−24.27 (0.7) b	−3.36 (1.1) b
5–15	DL	3.1 (0.8) a	11.2 (5.1) a	−21.60 (0.6) a	−0.41 (0.7) a
	SL	9.4 (5.2) b	22.4 (5.8) b	−24.20 (0.9) b	−2.17 (0.8) b
	AL	12.1 (2.1) b	27.3 (3.9) b	−24.04 (0.5) b	−3.38 (1.1) c
15–30	DL	1.8 (0.2) a	11.3 (2.2) a	−20.47 (1.1) a	−0.37 (0.8) a
	SL	6.2 (2.6) b	24.6 (6.3) b	−23.82 (0.8) b	−2.32 (0.9) b
	AL	10.7 (2.6) c	27.4 (3.9) b	−23.94 (1.5) b	−3.32 (1.1) c
30–50	DL	1.2 (0.5) a	8.5 (0.1) a	−21.56 (1.4) a	−0.64 (1.2) a
	SL	4.9 (2.4) ab	26.2 (6.7) b	−23.22 (0.9) a	−2.27 (1.0) ab
	AL	6.2 (3.0) b	30.0 (4.7) b	−22.70 (1.3) a	−3.33 (1.5) b
50–100	DL	1.4 (1.1) a	8.5 (2.4) a	−19.07 (0.2) a	−0.67 (1.8) a
	SL	4.2 (1.5) b	26.2 (8.9) b	−22.92 (1.0) b	−2.45 (1.4) a
	AL	3.7 (1.7) ab	30.4 (7.2) b	−22.59 (1.5) b	−2.93 (1.5) a

Values followed by the same letter in each column (for the same layer) are not significantly different, based on the LSD tests ($P < 0.05$)

significant differences, SOC content is larger (smaller) in the agricultural land than in the shrub land above (below) 50 cm. In particular, SOC content is significantly different over the 15–30 cm layer between the shrub land and agricultural land.

For all the depth, SIC content has the following order: desert land < shrub land < agricultural land. While statistical analyses only show a significant difference for the 0–5 cm, SIC content is generally higher (>10%) in the agricultural land relative to the shrub land. In addition, the variation in SIC (i.e., standard deviation) is also smaller in the agricultural land than in the shrub land, implying that the uncertainty (in terms of high SIC content) is small in the agricultural soils. The value of $\delta^{13}C$ in SOC is significantly higher in the desert land than in the shrub and agricultural lands; similarly, $\delta^{13}C$ in SIC follows an order: desert land > shrub land > agricultural land. Statistical analyses show that $\delta^{13}C$ in SIC is significantly different in the 5–30 cm between land uses.

Generally, the $\delta^{13}C$ value in LIC is close to 0, but more negative in PIC (Krull and Bray 2005; Liu et al. 1996). On average, the $\delta^{13}C$ in SIC of Yanqi Basin is −0.6, −2.2, and −3.4‰ for the desert land, shrub land, and agricultural land, respectively. The relatively depleted ^{13}C in SIC in the agricultural land indicates that there has been more PIC formed related to cropping.

It is well known that SOC is generally low in arid and semi-arid lands. The data collected in the Yanqi Basin show extremely low SOC density in the desert land (1.9 ± 1.4 kg C m^{-2} over the top 1 m) (Table 4). An earlier study by Feng et al.

(2002) showed close values, i.e., 2.3–2.4 kg C m^{-2} for the Tarim and Junggar Basins. SOC stocks (7.9–9.2 kg C m^{-2}) in the shrub and agricultural lands of Yanqi Basin are slightly lower than those reported by Wang et al. (2003), i.e., 10.9–12.1 kg C m^{-2} for shrub and agricultural lands in northwest China. But, the mean SOC stock of 8.6 kg C m^{-2} in the Yanqi Basin's shrub and agricultural lands is ~60% higher than the mean value (5.4 kg C m^{-2}) for the Xinjiang province reported by Li et al. (2007), but close to the national mean value for SOC (8.0 kg C m^{-2}) that was estimated using the China's second soil surveys data (Wu et al. 2003).

The SIC stock (9.7 ± 3.4 kg C m^{-2} over the 0–100 cm) in the desert land of the Yanqi Basin is slightly less than that in the Tarim Basin (10.8 kg C m^{-2}), but significantly higher than that in the Junggar Basin (3.2 kg C m^{-2}) (Feng et al. 2002). On the other hand, SIC stock is much higher (37–52 kg C m^{-2} over 0–100 cm) in Yanqi Basin's shrub and agricultural lands (Table 4), comparing with the average SIC density (12.6 kg C m^{-2}) for Xinjiang province reported by Li et al. (2007).

More than 80% of the total carbon stock is as the form of SIC in all the land use types of the Yanqi basin (Table 4), which implies the importance of SIC in the carbon storage in arid lands (Entry et al. 2004; Lal 2002; Wu et al. 2009). There is a greater amount of SIC accumulated in the subsoil of shrub and agricultural lands in the Yanqi basin, indicating the importance of deep soils for carbon storages in the arid and semi-arid lands (Wang et al. 2010).

3.1.4 Impacts of Land Use Changes on Soil Carbon Dynamics

It was well known that water limitation due to lower precipitation is a main factor limiting plant growth in arid and semi-arid lands, which is responsible for the low-SOC content in native lands. Earlier studies by Li et al. (2006) and Wang et al. (2011a) revealed an increase of SOC following the conversion of desert land to agricultural land in northwest China. Apparently, agricultural development with irrigation and fertilization can promote plant growth, which subsequently leads to an increase in root biomass and plant residue returned to the topsoil (Khan et al. 2009; Turner et al. 2011).

Land use change can have effects not only on SOC stock, but also on SIC stock in arid and semi-arid regions (Mikhailova and Post 2006; Wu et al. 2009; Zhang et al. 2010). Land use change (from shrub land to agricultural land) results in a large increase in SOC stock (43%) in the topsoil (0–30 cm), but SIC stock (56%) in the subsoil (30–100 cm) in the Yanqi Basin (Table 4). Similarly, Wu et al. (2009) reported an increase of SIC on irrigated croplands in northwest China. The study of Mikhailova and Post (2006) also showed that SIC stock (over the top 2 m) was significantly higher under continuous cultivation (242 Mg C ha^{-1}) than in native grassland (107 Mg C ha^{-1}). However, Wu et al. (2008)'s study reported that land use changes have different impacts on SIC at two sites in California, USA: Increased in the soil of Imperial Valley but decreased in the soil of San Joaquin, which might be caused by the differences in the irrigation waters (e.g., pH and salts contents and compositions).

Table 4 Means and percentages of both SOC and SIC stocks over 0–30 cm and 30–100 cm in the desert land (DL), shrub land (SL), and agricultural land (AL)

	0–30 cm				30–100 cm			
	SOC	SIC	SOC	SIC	SOC	SIC	SOC	SIC
	kg m^{-2}	kg m^{-2}	%	%	kg m^{-2}	kg m^{-2}	%	%
DL	0.86 ± 0.13	4.62 ± 1.30	7.49	40.24	0.94 ± 1.08	5.06 ± 2.67	8.19	44.08
SL	3.24 ± 1.29	9.79 ± 2.83	7.24	21.91	4.64 ± 2.65	27.01 ± 8.69	10.38	64.46
AL	4.61 ± 0.56	11.03 ± 1.51	9.01	21.55	4.62 ± 1.82	41.96 ± 5.23	9.02	60.41

4 Conclusions

(1) Vertical distribution is largely different between soil organic carbon and inorganic carbon, i.e., sharp decrease over depth in the former but various degrees of increasing trend in the latter.
(2) There are significant differences in soil organic and inorganic carbon contents between land use types in the Yanqi Basin. Both organic and inorganic carbon stocks are lowest in the desert land and highest in the agricultural land. The increase of organic carbon is found in the topsoil, but the increase of inorganic carbon is in the subsoils.
(3) On average, soil inorganic carbon stock in the Yanqi Basin counts more than 80% of the total soil stocks. The agricultural soils show significant depletion of ^{13}C in soil inorganic carbon, indicating a significant amount of secondary carbonate formed in association with cropping.

References

Boutton TW, Archer SR, Midwood AJ, Zitzer SF, Bol R (1998) $\delta^{13}C$ values of soil organic carbon and their use in documenting vegetation change in a subtropical savanna ecosystem. Geoderma 82:5–41. https://doi.org/10.1016/s0016-7061(97)00095-5

Cerling TE (1984) The stable isotopic composition of modern soil carbonate and its relationship to climate. Earth Planet Sci Lett 71:229–240. https://doi.org/10.1016/0012-821x(84)90089-x

De Vos B, Lettens S, Muys B, Deckers JA (2007) Walkley-Black analysis of forest soil organic carbon: recovery, limitations and uncertainty. Soil Use Manag 23:221–229. https://doi.org/10.1111/j.1475-2743.2007.00084.x

Dreimanis A (1962) Quantitative gasometric determination of calcite and dolomite by using Chittick apparatus. J Sediment Res 32:520

Entry JA, Sojka RE, Shewmaker GE (2004) Irrigation increases inorganic carbon in agricultural soils. Environ Manag 33:S309–S317. https://doi.org/10.1007/s00267-003-9140-3

FAO-UNESCO (1988) Soil map of the world: revised legend. World Soil Resources Report No 60 FAO, Rome

Feng Q, Endo KN, Cheng G (2002) Soil carbon in desertified land in relation to site characteristics. Geoderma 106:21–43

Fernandez I, Cadisch G (2003) Discrimination against ^{13}C during degradation of simple and complex substrates by two white rot fungi. Rapid Commun Mass Spectrom 17:2614–2620. https://doi.org/10.1002/rcm.1234

Gang H, Xue-yong Z, Yu-qiang L, Jian-yuan C (2012) Restoration of shrub communities elevates organic carbon in arid soils of northwestern China. Soil Biol Biochem 47:123–132. https://doi.org/10.1016/j.soilbio.2011.12.025

Heiri O, Lotter AF, Lemcke G (2001) Loss on ignition as a method for estimating organic and carbonate content in sediments: reproducibility and comparability of results. J Paleolimnol 25:101–110

Khan S, Hanjra MA, Mu J (2009) Water management and crop production for food security in China: a review. Agric Water Manag 96:349–360. https://doi.org/10.1016/j.agwat.2008.09.022

Krull EG, Bray SS (2005) Assessment of vegetation change and landscape variability by using stable carbon isotopes of soil organic matter. Aust J Bot 53:651–661. https://doi.org/10.1071/BT04124

Lal R (2002) Soil carbon sequestration in China through agricultural intensification, and restoration of degraded and desertified ecosystems. Land Degrad Dev 13:469–478. https://doi.org/10.1002/ldr.531

Lal R (2004) Carbon sequestration in dryland ecosystems. Environ Manag 33:528–544. https://doi.org/10.1007/s00267-003-9110-9

Li X, Li F, Rengel Z, Bhupinderpal S, Wang Z (2006) Cultivation effects on temporal changes of organic carbon and aggregate stability in desert soils of Hexi Corridor region in China. Soil and Tillage Res 91:22–29. https://doi.org/10.1016/j.still.2005.10.004

Li ZP, Han FX, Su Y, Zhang TL, Sun B, Monts DL, Plodinec MJ (2007) Assessment of soil organic and carbonate carbon storage in China. Geoderma 138:119–126. https://doi.org/10.1016/j.geoderma.2006.11.007

Liu B, Phillips FM, Campbell AR (1996) Stable carbon and oxygen isotopes of pedogenic carbonates, Ajo mountains, southern Arizona: implications for paleoenvironmental change. Palaeogeogr Palaeoclimatol Palaeoecol 124:233–246. https://doi.org/10.1016/0031-0182(95)00093-3

McPherson GR, Boutton TW, Midwood AJ (1993) Stable carbon isotope analysis of soil organic matter illustrates vegetation change at the grassland/woodland boundary in southeastern Arizona, USA. Oecologia 93:95–101. https://doi.org/10.1007/bf00321197

Mikhailova EA, Post CJ (2006) Effects of land use on soil inorganic carbon stocks in the Russian chernozem. J Environ Qual 35:1384–1388. https://doi.org/10.2134/jeq2005.0151

Nordt LC, Hallmark CT, Wilding LP, Boutton TW (1998) Quantifying pedogenic carbonate accumulations using stable carbon isotopes. Geoderma 82:115–136. https://doi.org/10.1016/s0016-7061(97)00099-2

Presley B (1975) A simple method for determining calcium carbonate in sediment samples. J Sediment Petrol 45:745–746

Sanderman J, Baldock J, Amundson R (2008) Dissolved organic carbon chemistry and dynamics in contrasting forest and grassland soils. Biogeochemistry 89:181–198. https://doi.org/10.1007/s10533-008-9211-x

Santisteban JI, Mediavilla R, López-Pamo E, Dabrio CJ, Zapata MBR, García MJG, Castaño S, Martínez-Alfaro PE (2004) Loss on ignition: a qualitative or quantitative method for organic matter and carbonate mineral content in sediments? J Paleolimnol 32:287–299. https://doi.org/10.1023/B:JOPL.0000042999.30131.5b

Schlesinger WH (1999) Carbon and agriculture—carbon sequestration in soils. Science 284:2095–2095

Sherrod LA, Dunn G, Peterson GA, Kolberg RL (2002) Inorganic carbon analysis by modified pressure-calcimeter method. Soil Sci Soc Am J 66:299–305. https://doi.org/10.2136/sssaj2002.2990

Su YZ, Wang XF, Yang R, Lee J (2010) Effects of sandy desertified land rehabilitation on soil carbon sequestration and aggregation in an arid region in China. J Environ Manage 91:2109–2116. https://doi.org/10.1016/j.jenvman.2009.12.014

Turner NC, Molyneux N, Yang S, Xiong Y-C, Siddique KHM (2011) Climate change in south-west Australia and north-west China: challenges and opportunities for crop production. Crop and Pasture Sci 62:445–456. https://doi.org/10.1071/CP10372

Vargas R, Carbone M, Reichstein M, Baldocchi D (2011) Frontiers and challenges in soil respiration research: from measurements to model-data integration. Biogeochemistry 102:1–13. https://doi.org/10.1007/s10533-010-9462-1

Walkley A, Black IA (1934) An examination of the Degtjareff method for determining soil organic matter and a proposed modification of the chromic acid titration method. Soil Sci 37:29–38

Wang X, Smethurst P, Herbert A (1996) Relationships between three measures of organic matter or carbon in soils of eucalypt plantations in Tasmania. Aust J Soil Res 34:545–553. https://doi.org/10.1071/SR9960545

Wang S, Tian H, Liu J, Pan S (2003) Pattern and change of soil organic carbon storage in China: 1960s–1980s. Tellus Ser B-Chem Phys Meteorol 55:416–427

Wang Y, Li Y, Ye X, Chu Y, Wang X (2010) Profile storage of organic/inorganic carbon in soil: from forest to desert. Sci Total Environ 408:1925–1931. https://doi.org/10.1016/j.scitotenv.2010.01.015

Wang K, Hua F, Ranab T, Hanjrac MA, Bo D, Huan L, Fenghua Z (2011a) Changes in soil carbon and nitrogen under long-term cotton plantations in China. J Agric Sci 149:497–505. https://doi.org/10.1017/S0021859611000049

Wang QR, Li YC, Wang Y (2011b) Optimizing the weight loss-on-ignition methodology to quantify organic and carbonate carbon of sediments from diverse sources. Environ Monit Assess 174:241–257. https://doi.org/10.1007/s10661-010-1454-z

Wang X, Wang J, Zhang J (2012) Comparisons of three methods for organic and inorganic carbon in calcareous soils of Northwestern China. PLoS ONE 7:e44334. https://doi.org/10.1371/journal.pone.0044334

Wu H, Guo Z, Peng C (2003) Distribution and storage of soil organic carbon in China. Global Biogeochem Cycles 17:1048. https://doi.org/10.1029/2001gb001844

Wu L, Wood Y, Jiang P, Li L, Pan G, Lu J, Chang AC, Enloe HA (2008) Carbon sequestration and dynamics of two irrigated agricultural soils in California. Soil Sci Soc Am J 72:808–814. https://doi.org/10.2136/sssaj2007.0074

Wu H, Guo Z, Gao Q, Peng C (2009) Distribution of soil inorganic carbon storage and its changes due to agricultural land use activity in China. Agric Ecosyst Environ 129:413–421. https://doi.org/10.1016/j.agee.2008.10.020

Yang YH, Mohammat A, Feng JM, Zhou R, Fang JY (2007) Storage, patterns and environmental controls of soil organic carbon in China. Biogeochemistry 84:131–141. https://doi.org/10.1007/s10533-007-9109-z

Zhang N, He X-D, Gao Y-B, Li Y-H, Wang H-T, Ma D, Zhang R, Yang S (2010) Pedogenic carbonate and soil dehydrogenase activity in response to soil organic matter in *Artemisia ordosica* community. Pedosphere 20:229–235. https://doi.org/10.1016/s1002-0160(10)60010-0

Distribution of Pedogenic Carbonate and Relationship with Soil Organic Carbon in Yanqi Basin

Xiujun Wang, Jiaping Wang and Junyi Wang

Abstract Studying of pedogenic carbonate (PIC) has been lacking despite its importance in carbon sequestration. Using isotopic approach, vertical distribution of PIC over 0–100 cm is assessed for agricultural and native lands. In general, PIC stock is significantly higher in the agricultural land than in the native lands. There is a strong correlation ($P < 0.001$) between PIC stock and SOC stock, implying that an increase of 1 kg C m^{-2} in SOC stock may lead to an increase of 1.9 kg C m^{-2} in PIC formation. Conversion of native lands to agricultural land has led to an increase in both organic carbon and PIC in soil profiles of Yanqi Basin.

1 Introduction

Soil carbonate or inorganic carbon (SIC), primarily calcium carbonate and magnesium carbonate, is commonly seen in arid and semi-arid regions. The SIC pool consists of two components: lithogenic carbonate (LIC) and pedogenic carbonate (PIC). In general, LIC originates largely as detritus from the parent materials. But, PIC, also termed as secondary carbonate, is formed mainly in two ways: (1) by dissolution and re-precipitation of LIC, and (2) through dissolution of carbon dioxide (CO_2) into HCO_3^-, then precipitation with Ca^{2+} and/or Mg^{2+} that originate from non-LIC minerals (e.g., weathering silicate, dust, and fertilizers). Thus, the formation of PIC from the non-LIC minerals can lead to carbon sequestration (Monger and Gallegos 2000).

There is a large uncertainty in the estimation of global PIC stock, which may be partly due to the difficulty of separating PIC and LIC (Eswaran et al. 2000). There are a few approaches that may be applied to separate LIC and PIC. Because LIC and

---

X. Wang (✉) · J. Wang
College of Global Change and Earth System Science,
Beijing Normal University, Beijing 100875, China
e-mail: xwang@bnu.edu.cn

J. Wang
College of Agriculture, Shihezi University, Shihezi, Xinjiang 832000, China

PIC have distinct isotopic values that are related to different sources of carbonate, isotope technique has been used to quantify the proportions of LIC and PIC in soils (Breecker et al. 2009; Landi et al. 2003; Rabenhorst et al. 1984). In general, the $\delta^{13}C$ value in LIC is close to zero, but more negative in PIC (Krull and Bray 2005; Liu et al. 1996).

Although it is a common phenomenon that soil organic carbon (SOC) is extremely low in arid and semi-arid lands, there is widespread evidence that SIC stock is 2–10 times as high as SOC stock in soil profiles (Scharpenseel et al. 2000; Wang et al. 2010). Recent studies have shown that in the Northern China, higher SIC stock is often associated with higher SOC stock (Wang et al. 2014), and SIC accumulation is mainly in the subsoils, which imply enhancement of PIC in soil profiles. However, little is known about the dynamics of PIC in the arid and semi-arid lands.

While there are indications of PIC having potential for carbon sequestration and climate mitigation (Eshel et al. 2007; Lal and Kimble 2000; Manning 2008), there have been limited studies (Scharpenseel et al. 2000; Schlesinger 1982) of quantifying carbon sequestration as PIC for the vast arid and semi-arid regions. On the one hand, some studies showed that PIC accumulation was extremely low (<3 g C m^{-2} y^{-1}) in the arid and semi-arid regions of Canada, USA, and New Zealand (Landi et al. 2003; Scharpenseel et al. 2000). On the other hand, limited studies demonstrated significant PIC accumulation (>10 g C m^{-2} y^{-1}) in the arid and semi-arid lands of Northern China (Pan and Guo 2000; Wang et al. 2014). Apparently, more studies are needed to evaluate the PIC dynamics under various land uses across the arid and semi-arid regions.

A study was carried out in the Yanqi Basin to estimate the vertical distribution of PIC, and to investigate the relationship between PIC and SOC stocks. Soil samples were collected from 21 profiles: three in desert land, nine in shrubland, and nine in cropland. This work was a part of integrated studies that aim to improve the understanding of terrestrial carbon cycle in the arid regions.

2 Materials and Methods

2.1 Soil Sampling and Analyses

Soil samples were collected from representative soil profiles in the Yanqi Basin during August and November, 2010. There were three desert sites, and nine sites each from shrubland and cropland. Soil samples were grounded to pass a 0.25-mm screen for analyses of SOC and SIC contents and the stable isotopic compositions. In this study, we also included SOC and SIC data of four extra profiles from cropland.

Contents of total carbon and SOC were measured using a CNHS–O analyzer. Briefly, ~1 g soil was combusted at 1020 °C with a constant helium flow carrying pure oxygen to ensure completed oxidation of organic materials. Production of CO_2 was determined by a thermal conductivity detector. Soil total carbon (STC) and SOC were determined without and with pretreatment of HCl, respectively. SIC was calculated as the difference between STC and SOC. Stable carbon isotope was measured using an isotope ratio mass spectrometer (Delta Plus XP, Thermo Finnigan MAT, Germany). Detailed procedures of sampling and analyses were reported by Wang et al. (2015a).

2.2 Estimation of Pedogenic Carbonate

Following Landi et al. (2003) and Wang et al. (2014), PIC was calculated as:

$$PIC = \frac{\delta^{13}C_{SIC} - \delta^{13}C_{PM}}{\delta^{13}C_{PIC} - \delta^{13}C_{PM}} SIC \quad (1)$$

where $\delta^{13}C_{SIC}$, $\delta^{13}C_{PM}$, and $\delta^{13}C_{PIC}$ were the stable ^{13}C in carbonate for the bulk SIC, parent material, and pure PIC, respectively. The value of $\delta^{13}C_{PM}$ was set as 0.58‰, the largest $\delta^{13}C$ value that was obtained from a sample in the desert land. Because it was not possible to collect PIC samples, we calculated $\delta^{13}C_{PIC}$ using the following method:

$$\delta^{13}C_{PIC} = \delta^{13}C_{SOC} + 14.9 \quad (2)$$

where $\delta^{13}C_{SOC}$ is the stable ^{13}C in SOC. The value of 14.9 represents the sum of the isotopic fractionation for CO_2 diffusion and carbonate precipitation (Cerling 1984; Cerling et al. 1989, 1991).

There is evidence that atmospheric CO_2 may be transferred into soil pores under low rates of soil respiration (Breecker et al. 2009; Cerling 1984; Stevenson et al. 2005), which would alter the isotopic composition for soil CO_2. To assess the effect of atmospheric CO_2 mixing, we calculated the stable ^{13}C in soil CO_2 as:

$$\delta^{13}C_{SC} = (1 - \lambda)\delta^{13}C_{RC} + \lambda\delta^{13}C_{Air} \quad (3)$$

$$\delta^{13}C_{RC} = \delta^{13}C_{SOC} + 4.4 \quad (4)$$

where the terms $\delta^{13}C_{RC}$ and $\delta^{13}C_{Air}$ are the stable ^{13}C in the respired and atmospheric CO_2, respectively, and λ the contributions of atmospheric CO_2 to soil CO_2. We applied this process to the upper 50 cm at those sites that contain less than 7 g kg^{-1} SOC in topsoil (0–30 cm). We set $\delta^{13}C_{Air}$ to −8‰, and λ value to 1, 0.9, 0.7, and 0.5 for the 0–5, 5–15, 15–30, and 30–50 cm, respectively.

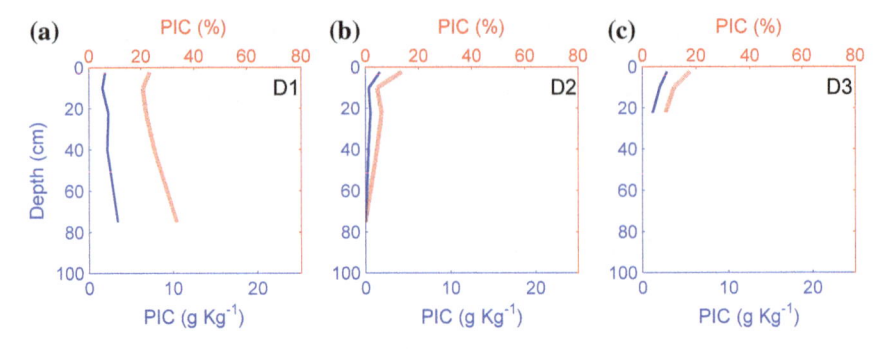

Fig. 1 Vertical distributions of PIC in the desert land

3 Results and Discussions

3.1 Basic Properties for the Sampling Sites

Soil pH ranges from 7.8 to 9.1 and shows no significant difference among land use types (Table 1). Both exchangeable Ca and Mg contents are generally high (>11.7 and >0.22 g kg^{-1}, respectively) in the surface soils, except in the desert lands that contain <8 g kg^{-1} Ca and <0.08 g. Contents of exchangeable Ca and Mg, and SOC and SIC follow the same order: agricultural land > shrubland > desert land. The $\delta^{13}C_{SOC}$ value in the surface soil ranges narrowly, from −21.7 to −24.8‰, whereas the $\delta^{13}C$ value in root shows a range from −23.4 to −28.3‰ for C3 native plants, and from −11.4 to −12.6‰ for maize.

3.2 Vertical Distribution of PIC

The percentage of PIC is low in the desert land, i.e., <20% at two of the three sites (Fig. 1). Clearly, PIC content is extremely low (<4 g kg^{-1}), with little vertical variation in all profiles. Figure 2 shows that PIC percentage varies between ~20 and ~40%, and PIC content from ~3 to ~13 g kg^{-1} in the shrubland. There are weak-to-strong vertical variations in both the percentage and content of PIC, with various patterns. The S1 and S9 sites are much different from the rest, with the strongest in the former and the weakest in the latter. In particular, the S1 shows the lowest values (e.g., PIC < 20% or ~4 g kg^{-1}) near the surface but the highest PIC content (~12 g kg^{-1}) in the subsurface. On the other hand, PIC is almost uniformly distributed in the profile at the S9 site, showing very low values with a small range (approximately 3–5 g kg^{-1}).

There are large differences in the magnitude and pattern of vertical distribution in PIC in the agricultural land (Fig. 3). While ~50% of sites have high PIC percentage (>40%) and content (>10 g kg^{-1}), there are two sites showing very low values

Table 1 Basic chemical properties in surface soils and isotope in roots from desert land (DL), shrubland (SL), and agricultural land (AL) (after Wang et al. 2015b)

Vegetation	Type	pH	Ca (g kg^{-1})	Mg (g kg^{-1})	SOC (g kg^{-1})	SIC (g kg^{-1})	$\delta^{13}C_{Root}$ (‰)	$\delta^{13}C_{SOC}$ (‰)
Populus tomentosa Carr	DL	8.30	3.92	0.04	3.7	10.3	n.d.	−21.32
Artemisia	DL	8.41	7.71	0.07	3.4	16.7	−23.36	−21.71
Sophora alopecuroides Linn	SL	8.30	14.2	0.78	5.7	25.4	−25.23	−24.76
Leguminosae	SL	8.72	11.7	0.83	4.9	24.6	−25.67	−23.99
Achnatherum splendens (Trin.) Nevski	SL	8.42	12.5	0.98	16.8	22.5	−24.97	−25.28
Leguminosae	SL	8.44	13.4	0.73	14.9	23.2	−23.67	−25.09
Halostachys caspica (Bieb.) C. A Mey	SL	8.26	13.7	0.66	14.3	17.3	−23.15	−24.51
Phragmites australis	SL	7.94	16.6	0.33	9.3	27.8	−25.05	−24.40
Achnatherum splendens (Trin.) Nevski	SL	9.07	15.3	0.29	5.7	30.1	−25.60	−24.09
Halostachys caspica(Bieb.) C. A Mey	SL	8.23	13.3	0.59	7.5	22.3	−28.29	−23.52
Tamarix ramosissima Ledeb.	SL	8.31	9.87	0.22	3.4	10.0	−26.82	−22.72
Capsicum annuum Linn	AL	8.21	13.2	0.23	9.5	25.4	−26.84	−24.06
Zea mays	AL	8.43	13.5	2.40	14.0	31.3	−12.57	−25.17
Gossypium spp.	AL	8.18	12.7	1.46	14.9	27.7	−24.70	−24.41
Beta vulgaris	AL	8.23	13.6	0.71	10.7	27.3	n.d.	−24.03
Capsicum annuum Linn	AL	7.91	14.5	1.08	9.5	32.6	−24.04	−23.65
Zea mays	AL	8.22	15.8	0.29	11.6	30.5	−11.90	−23.47
Helianthus annuus	AL	8.44	15.2	0.81	11.7	26.1	−26.88	−24.42
Zea mays	AL	7.84	13.2	0.62	11.6	19.7	−11.41	−24.23
Brassica campestris L.	AL	8.35	14.7	1.46	12.6	24.9	−26.09	−23.23

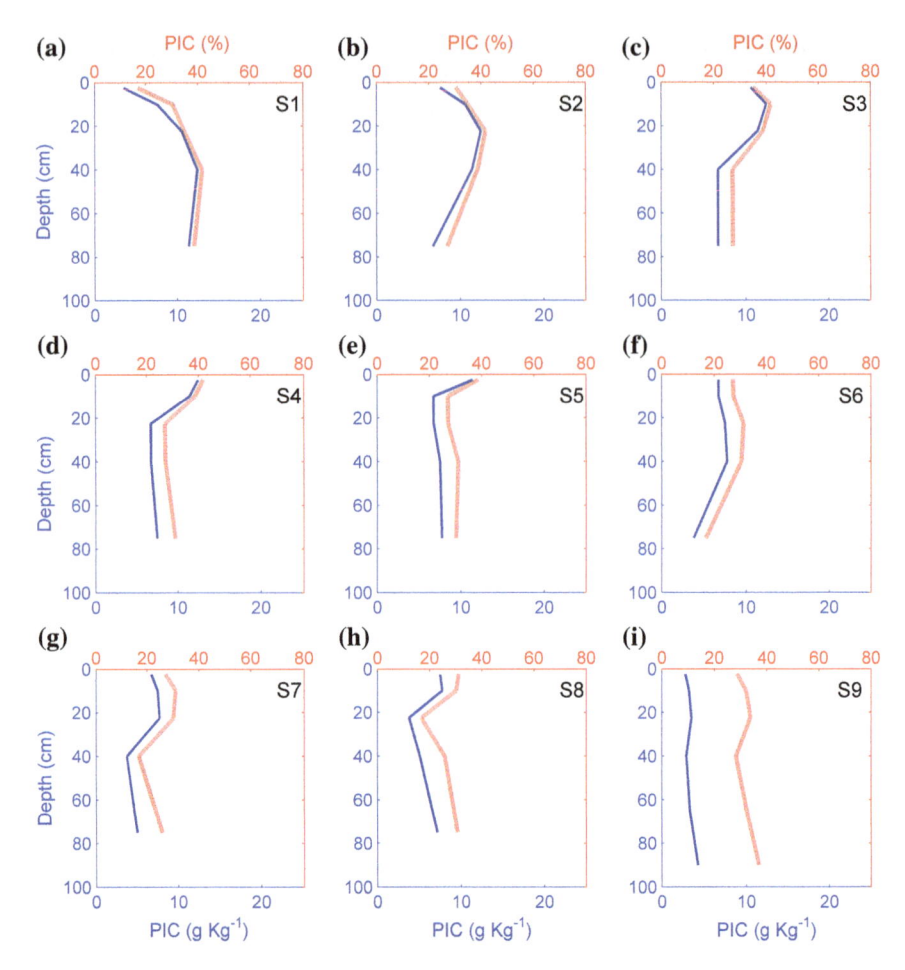

Fig. 2 Vertical distributions of PIC in the shrubland

(<5 g kg^{-1}) either in the topsoil or in the subsoil. The majority of the sites reveal significant enhancement of PIC in the subsoils. In particular, PIC content reaches 15 g kg^{-1} in the subsurface at four sites (A2, A3, A7, and A9), which is significantly higher than those in the shrubland.

3.3 Relationship Between Carbonate and SOC Stocks

It appears that there are large differences in the magnitude and vertical variation of PIC within and between land use types. To better understand the dynamics of carbonate, various correlation analyses were carried out. Using all data from all the

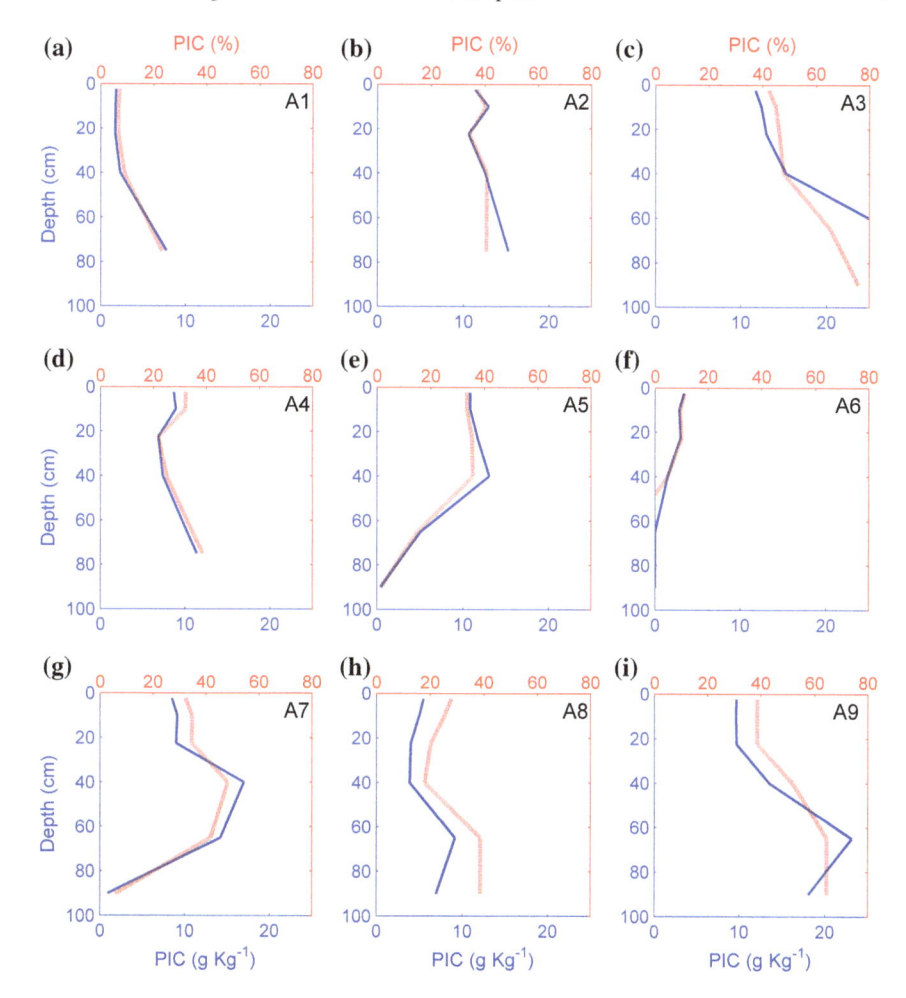

Fig. 3 Vertical distribution of PIC in the agricultural land

layers yields a weak relationship ($r^2 = 0.06$, $P > 0.05$) between SIC content and SOC content (Fig. 4), but a significantly negative correlation ($r^2 = 0.55$, $P < 0.01$) between SIC content and its isotope ($\delta^{13}C_{SIC}$) (Fig. 5). Apparently, high SIC levels (>20 g kg^{-1}) are corresponding with more negative $\delta^{13}C_{SIC}$ (-3 to $-7‰$), which indicates that a significant proportion of SIC is as PIC in the soils with high SIC contents.

To further evaluate the relationship between carbonate and SOC, each profile's SOC, SIC, and PIC stocks are calculated over the 0–100 cm. As shown in Fig. 6, there is a significantly positive correlation ($r^2 = 0.65$, $P < 0.01$) between SIC and SOC stocks although there is no significant relationship between the SOC and SIC stocks in soils collected from the cropland. The estimated PIC stock has a range of 0–34 kg C m^{-2} over the 0–100 cm, which is highly ($r^2 = 0.56$, $P < 0.01$) correlated

Fig. 4 Relationship between SIC content and SOC content using all data

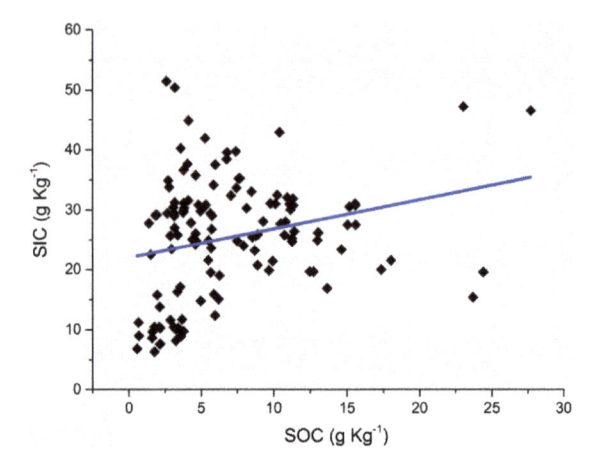

Fig. 5 Relationship between SIC content and its isotopic value using all data (redrawn from Wang et al. 2015b)

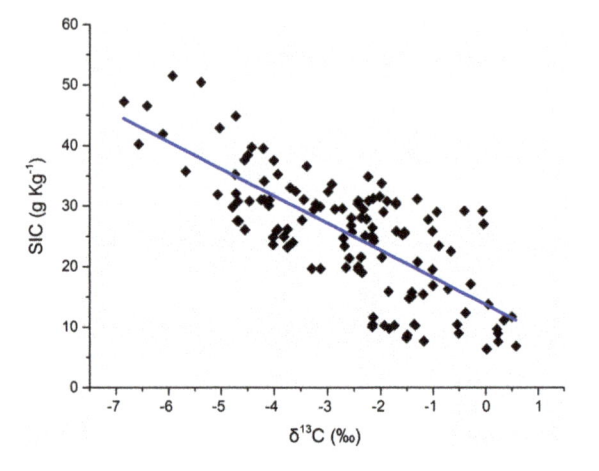

with SOC stock (Fig. 7). It appears that an increase of 1 kg C m^{-2} in SOC stock may lead to an increase of 1.9 kg C m^{-2} in PIC formation in Yanqi Basin.

An earlier study reported a negative relationship between SIC and SOC in the semi-arid regions of northwest China (Pan and Guo 2000). However, later studies reported a positive correlation between the SOC and SIC stocks in China's arid regions, which include near the eastern boundary of Tengger Desert, Inner Mongolia (Zhang et al. 2010), at the edge of Badan Jaran Desert, Gansu (Su et al. 2010), and in the North China Plain (Shi et al. 2017). These inconsistent findings may imply that the relationship between SOC and SIC is complex because of various processes involved with the accumulation and transformation, and decoupling of these processes over the space and time (Zheng et al. 2011).

Fig. 6 Relationship between SIC and SOC stocks (over the 0–100 cm) in desert land (stars), semi-desert land (diamonds), shrubland (triangles), and cropland (circles) (redrawn from Wang et al. 2015b)

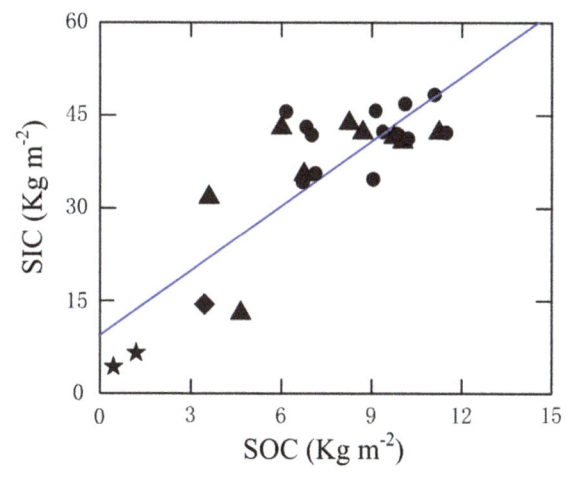

Fig. 7 Relationship between PIC and SOC stocks (over the 0–100 cm) (redrawn from Wang et al. 2015b). Symbols are the same as in Fig. 6

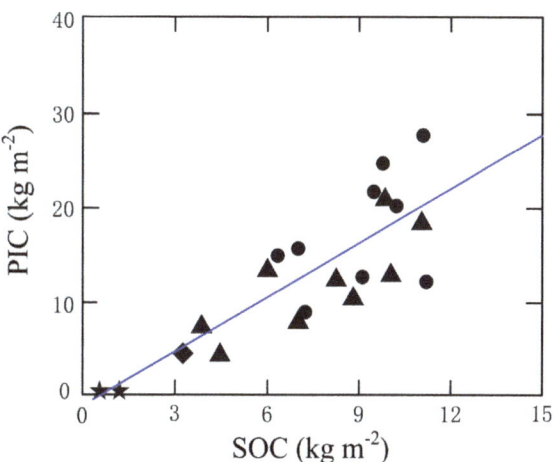

3.4 Soil Carbon Accumulation Rates in the Yanqi Basin

Table 2 illustrates an increase in almost all soil carbon stocks in the agricultural land relative to the shrubland. On average, SOC stock in the top 0–30 cm increases from 3.2 kg C m^{-2} in the shrubland to 4.6 kg C m^{-2} in the cropland. On the other hand, the increase of SIC stock is much larger in the subsoil (3.9 kg C m^{-2}) than in the topsoil (1.2 kg C m^{-2}) between the cropland and shrubland. Similarly, the increase of PIC is mainly found in the subsoil. Statistical analyses indicated a significant increase of SOC in the topsoil ($P < 0.01$), but SIC and PIC in the subsoil ($P < 0.05$) in the cropland. On average, the SOC, SIC, and PIC stocks over the 0–100 cm increased by 1.4, 5.1, and 5.2 kg C m^{-2}, respectively, as a result of land use change.

Table 2 Means (standard deviations) of SOC, SIC, PIC, and LIC stocks (kg C m^{-2}) for agricultural land (n = 9) and shrubland (n = 9) and accumulation rate (g C m^{-2} yr^{-1}) in agricultural land (after Wang et al. 2015b)

Stock	Cropland		Shrubland		Accumulation rate[a]	
	0–30	30–100	0–30	30–100	0–30	30–100
SOC	4.6 (0.6)	4.6 (1.8)	3.2 (1.3)	4.6 (1.7)	23.3	0.0
SIC	11.0 (1.5)	30.9 (4.3)	9.8 (2.8)	27.0 (6.9)	20.0	65.0
PIC	4.4 (1.3)	14.2 (6.7)	3.8 (1.6)	9.6 (5.2)	10.0	76.7
LIC	6.6 (2.2)	16.7 (5.9)	6.0 (2.3)	17.2 (7.0)	10.0	−11.7
Total C	15.6	35.5	13.0	31.6	43.3	65.0

[a]Accumulation rate is calculated as the difference divided by 60 years

There have been a number of studies on the magnitudes and spatial distributions of SIC at regional scales in the Northern China (Li et al. 2007; Wu et al. 2009). Yet, detailed analyses on SIC accumulation are limited. A study shows a decrease (at a rate of 26.8 g C m^{-2} year^{-1}) in SIC stock over the top 10 cm in China's grassland, particularly in the places with stronger soil acidification (Yang et al. 2012). However, data from the Northern China's cropland demonstrate a significant increase (101–202 g C m^{-2} year^{-1}) in the SIC stock over the 0–100 cm, with a larger increase found under the long-term application of organic materials (Wang et al. 2014).

Given that the cropland in the Yanqi Basin was converted from shrubland 60 years ago, one may estimate the accumulation rates of soil carbon stocks for the cropland by assuming that the differences between the cropland and shrubland are a result of land use changes over the past 60 years. This approach yields an accumulation rate of 20 g C m^{-2} year^{-1} for SIC over the 0–30 cm for the cropland in the Yanqi Basin (Table 2), which is slightly lower than that for SOC (23 g C m^{-2} year^{-1}). However, the accumulation rate of SIC is 85 g C m^{-2} year^{-1} over the 0–100 cm, with larger accumulation of SIC in the subsoil.

The estimated PIC accumulation rate reaches 87 g C m^{-2} year^{-1} in the cropland of Yanqi Basin, which is significantly higher than previously reported rates, including those (<3 g C m^{-2} y^{-1}) for Canada, USA, and New Zealand (Landi et al. 2003; Scharpenseel et al. 2000), and the earlier reported rates (10–40 g C m^{-2} y^{-1}) for the Aridisols in the northwest China (Pan and Guo 2000), but comparable with the recent reported rate for the cropland of Northern China (Wang et al. 2014). The large discrepancy is probably attributed to the differences in various factors between these regions, e.g., Ca^{2+} and Mg^{2+} availability (Monger and Gallegos 2000; Scharpenseel et al. 2000). Soils in the Mojave Desert and semi-arid region of Canada may be limited by calcium (Hirmas and Graham 2011; Landi et al. 2003; Monger and Gallegos 2000), whereas the croplands in the Northern China may have various sources of Ca and/or Mg, including fertilizers, dust, irrigation water, and weathering of calcium/magnesium silicate minerals (Wang et al. 2014). The extractable Ca in the Yanqi Basin is significantly higher in the cropland and shrubland than in the desert land (Table 1), indicating that groundwater may play a role in supplying Ca.

3.5 Uncertainty in Estimation of PIC by Isotopic Approach

Pedogenic carbonate often forms in equilibrium with soil CO_2 that primarily generates from SOC decomposition and root respiration. Thus, soil CO_2 would have isotopic signatures of SOC, C3 plant and C4 plant, and isotope technique may be used to determine the contributions of different sources of CO_2 and proportion of PIC in total SIC (Eshel et al. 2007; Landi et al. 2003; Mermut et al. 2000). However, there may be uncertainty in the estimated PIC by the isotopic technique due to the lack of relative parameters; thus, various assumptions made during the calculation. In particular, there are limited observations in terms of magnitude and variability of CO_2 production in soil profile.

While some studies indicate that CO_2 production due to root respiration may exceed that by SOC decomposition in arid and semi-arid ecosystems (Breecker et al. 2012; Li et al. 2011), one may also expect that SOC decomposition is the main contributor for CO_2 production during non-growing season or under little vegetation. In addition, there may be large differences in the vertical variation between SOC decomposition and root respiration, and the former can be dominant below the root zone.

There may be uncertainty in the estimated PIC due to the choices for $\delta^{13}C_{PM}$ and assumption for the contribution of SOC to the respired CO_2. Sensitivity studies indicate that applying a smaller value for $\delta^{13}C_{PM}$ results in a considerable decrease (>10%) in the percentage of PIC, whereas changing the contribution of SOC to CO_2 production has little effect on PIC estimation (Wang et al. 2015b). On the other hand, previous studies show that atmospheric CO_2 may be transferred into soil pores under low rate of soil respiration (Breecker et al. 2009; Cerling 1984; Stevenson et al. 2005), which would alter the isotopic composition for soil CO_2. Primarily analysis indicated that leaving this process out in the equation might lead to underestimation of PIC percentage by 20% in the shrubland, and thus over-estimation of the difference in PIC stock between the cropland and shrubland (Wang et al. 2015b). Apparently, to address the intriguing issue of CO_2 uptake in arid lands, further studies are needed to quantify the magnitudes of major carbon pools (e.g., SOC, SIC, and PIC), transformations and fluxes among these pools, and the dynamics of soil CO_2 in various ecosystems over multi-spatial and temporal scales.

References

Breecker DO, Sharp ZD, McFadden LD (2009) Seasonal bias in the formation and stable isotopic composition of pedogenic carbonate in modem soils from central New Mexico, USA. Geol Soc Am Bull 121:630–640. https://doi.org/10.1130/b26413.1

Breecker DO, McFadden LD, Sharp ZD, Martinez M, Litvak ME (2012) Deep autotrophic soil respiration in shrubland and woodland ecosystems in Central New Mexico. Ecosystems 15:83–96. https://doi.org/10.1007/s10021-011-9495-x

Cerling TE (1984) The stable isotopic composition of modern soil carbonate and its relationship to climate. Earth and Planetary Science Letters 71:229–240. https://doi.org/10.1016/0012-821x(84)90089-x

Cerling TE, Quade J, Wang Y, Bowman JR (1989) Carbon isotopes in soils and paleosols as ecology and paleoecology indicators. Nature 341:138–139. https://doi.org/10.1038/341138a0

Cerling TE, Solomon DK, Quade J, Bowman JR (1991) On the isotopic composition of carbon in soil carbon-dioxide. Geochim Cosmochim Acta 55:3403–3405. https://doi.org/10.1016/0016-7037(91)90498-t

Eshel G, Fine P, Singer MJ (2007) Total soil carbon and water quality: an implication for carbon sequestration. Soil Sci Soc Am J 71:397–405. https://doi.org/10.2136/sssaj2006.0061

Eswaran H, Reich PF, Kimble JM, Beinroth FH, Padmanabhan E, Moncharoen P (2000) Global carbon stocks. In: L R., JM Kimble, H Eswaran, BA Stewart (eds) Global climate change and pedogenic carbonates. Lewis Publishers, Boca Raton, FL

Hirmas DR, Graham RC (2011) Pedogenesis and soil-geomorphic relationships in an arid mountain range, Mojave desert, California. Soil Sci Soc Am J 75:192–206. https://doi.org/10.2136/sssaj2010.0152

Krull EG, Bray SS (2005) Assessment of vegetation change and landscape variability by using stable carbon isotopes of soil organic matter. Aust J Bot 53:651–661. https://doi.org/10.1071/BT04124

Lal R, Kimble JM (2000) Pedogenic carbonate and the global carbon cycle. In: Lal R, Kimble JM, Eswaran H, Stewart BA (eds) Global climate change and pedogenic carbonate. CRC Press, Boca Raton, FL, USA

Landi A, Mermut AR, Anderson DW (2003) Origin and rate of pedogenic carbonate accumulation in Saskatchewan soils, Canada. Geoderma 117:143–156

Li ZP, Han FX, Su Y, Zhang TL, Sun B, Monts DL, Plodinec MJ (2007) Assessment of soil organic and carbonate carbon storage in China. Geoderma 138:119–126. https://doi.org/10.1016/j.geoderma.2006.11.007

Li ZG, Wang XJ, Zhang RH, Zhang J, Tian CY (2011) Contrasting diurnal variations in soil organic carbon decomposition and root respiration due to a hysteresis effect with soil temperature in a *Gossypium* s. (cotton) plantation. Plant Soil 343:347–355. https://doi.org/10.1007/s11104-011-0722-1

Liu B, Phillips FM, Campbell AR (1996) Stable carbon and oxygen isotopes of pedogenic carbonates, Ajo mountains, southern Arizona: implications for paleoenvironmental change. Palaeogeogr Palaeoclimatol Palaeoecol 124:233–246. https://doi.org/10.1016/0031-0182(95)00093-3

Manning DAC (2008) Biological enhancement of soil carbonate precipitation: passive removal of atmospheric CO_2. Mineral Mag 72:639–649. https://doi.org/10.1180/minmag.2008.072.2.639

Mermut AR, Amundson R, Cerling TE (2000) The use of stable isotopes in studying carbonate dynamics in soils. In: Lal R, Kimble JM, Eswaran H, Stewart BA (eds) Global climate change and pedogenic carbonates. CRC Press, Boca Raton, FL USA

Monger HC, Gallegos RA (2000) Biotic and abiotic processes and rates of pedogenic carbonate accumulation in the southwestern United States-relationship to atmospheric CO_2 sequestration. In: Lal R, Kimble JM, Eswaran H, Stewart BA (eds) Global climate change and pedogenic carbonate. CRC Press, Boca Raton, FL, USA

Pan GX, Guo T (2000) Pedogenic carbonate of aridic soils in China and its signficance in carbon sequestration in terrretrial systems. In: Lal R, Kimble JM, Eswaran H, Stewart BA (eds) Global climate change and pedogenic carbonates. CRC Press, Boca Raton, FL, USA

Rabenhorst MC, Wilding LP, West LT (1984) Identification of pedogenic carbonates using stable carbon isotope and microfabric analyses1. Soil Sci Soc Am J 48:125–132. https://doi.org/10.2136/sssaj1984.03615995004800010023x

Scharpenseel HW, Mtimet A, Freytag J (2000) Soil inorganic carbon and global change. In: Lal R, Kimble JM, Eswaran H, Stewart BA (eds) Global climate change and pedogenic carbonates. CRC Press, Boca Raton, FL

Schlesinger WH (1982) Carbon storage in the caliche of arid soils—a case study from Arizona. Soil Sci 133:247–255. https://doi.org/10.1097/00010694-198204000-00008

Shi HJ, Wang XJ, Zhao YJ, Xu MG, Li DW, Guo Y (2017) Relationship between soil inorganic carbon and organic carbon in the wheat-maize cropland of the North China plain. Plant Soil 418:423–436. https://doi.org/10.1007/s11104-017-3310-1

Stevenson BA, Kelly EF, McDonald EV, Busacca AJ (2005) The stable carbon isotope composition of soil organic carbon and pedogenic carbonates along a bioclimatic gradient in the Palouse region, Washington State, USA. Geoderma 124:37–47. https://doi.org/10.1016/j.geoderma.2004.03.006

Su YZ, Wang XF, Yang R, Lee J (2010) Effects of sandy desertified land rehabilitation on soil carbon sequestration and aggregation in an arid region in China. J Environ Manage 91:2109–2116. https://doi.org/10.1016/j.jenvman.2009.12.014

Wang Y, Li Y, Ye X, Chu Y, Wang X (2010) Profile storage of organic/inorganic carbon in soil: from forest to desert. Sci Total Environ 408:1925–1931. https://doi.org/10.1016/j.scitotenv.2010.01.015

Wang XJ, Xu MG, Wang JP, Zhang WJ, Yang XY, Huang SM, Liu H (2014) Fertilization enhancing carbon sequestration as carbonate in arid cropland: assessments of long-term experiments in northern China. Plant Soil. https://doi.org/10.1007/s11104-11014-12077-x

Wang JP, Wang XJ, Zhang J, Zhao CY (2015a) Soil organic and inorganic carbon and stable carbon isotopes in the Yanqi Basin of northwestern China. Eur J Soil Sci 66:95–103. https://doi.org/10.1111/ejss.12188

Wang XJ, Wang JP, Xu MG, Zhang WJ, Fan TL, Zhang J (2015b) Carbon accumulation in arid croplands of northwest China: pedogenic carbonate exceeding organic carbon. Sci Rep 5:11439. https://doi.org/10.1038/srep11439

Wu HB, Guo ZT, Gao Q, Peng CH (2009) Distribution of soil inorganic carbon storage and its changes due to agricultural land use activity in China. Agr Ecosyst Environ 129:413–421. https://doi.org/10.1016/j.agee.2008.10.020

Yang YH, Fang JY, Ji CJ, Ma WH, Mohammat A, Wang SF, Wang SP, Datta A, Robinson D, Smith P (2012) Widespread decreases in topsoil inorganic carbon stocks across China's grasslands during 1980s–2000s. Glob Change Biol 18:3672–3680. https://doi.org/10.1111/gcb.12025

Zhang N, He XD, Gao YB, Li YH, Wang HT, Ma D, Zhang R, Yang S (2010) Pedogenic carbonate and soil dehydrogenase activity in response to soil organic matter in Artemisia ordosica community. Pedosphere 20:229–235. https://doi.org/10.1016/s1002-0160(10)60010-0

Zheng J, Cheng K, Pan G, Pete S, Li L, Zhang X, Zheng J, Han X, Du Y (2011) Perspectives on studies on soil carbon stocks and the carbon sequestration potential of China. Chin Sci Bull 56:3748–3758. https://doi.org/10.1007/s11434-011-4693-7

Spatial Distribution of Organic Carbon in Surface Sediment of Bosten Lake

Zhitong Yu, Xiujun Wang and Hang Fan

Abstract Lake sediment is a vital carbon reservoir, which is affected by biogeochemical and hydrological processes in the watershed. To study the dynamics of organic carbon in surface sediment of the Bosten Lake, we analyzed total organic carbon (TOC) and its stable carbon isotopic composition ($\delta^{13}C_{org}$), total nitrogen (TN), and grain size in the surface (0–2 cm) sediment. Our data showed that there was a large spatial variability in both TOC (2.1–4.2%) and $\delta^{13}C_{org}$ (−26.65 to −24.13‰) in surface sediment of the Bosten Lake. By using a three end member mixing model, we estimated that 54–90% of TOC was from autochthonous source. We found higher TOC concentration (>3.7%) near the mouth of the Kaidu River, in the central-north section and in the east section, which was attributable to autochthonous, autochthonous plus allochthonous, and allochthonous sources, respectively. The lowest TOC was seen in the mid-west section, which might mainly be due to the high kinetic energy levels. Our analyses suggested that the magnitude and spatial distribution of TOC in the surface sediment of Bosten Lake were influenced by complex processes and regulated by multiple factors.

1 Introduction

Lakes, rivers and other inland water bodies, as unique components on the Earth, are strongly influenced by the watershed's biogoechemical processes. Despite of the relatively small coverage (Downing et al. 2006), lakes play a crucial role in the biogeochemical cycle of the terrestrial ecosystems because of the high sedimentation

---

Z. Yu (✉) · X. Wang · H. Fan
College of Global Change and Earth System Science, Beijing Normal University, 100875 Beijing, China
e-mail: yuzhitong@bnu.edu.cn

Z. Yu
Xinjiang Institute of Ecology and Geography, Chinese Academy of Sciences, 830011 Urumqi, China

© Springer Nature Singapore Pte Ltd. 2018
X. Wang et al. (eds.), *Carbon Cycle in the Changing Arid Land of China*,
Springer Earth System Sciences, https://doi.org/10.1007/978-981-10-7022-8_8

rates and a large amount of carbon burial in the lakes's sediments (Battin et al. 2009; Dean and Gorham 1998; Tranvik et al. 2009).

Total organic carbon (TOC) exhibits large spatial differences in lake sediment across the world, including the North America (Dean and Gorham 1998), West Europe (Bechtel and Schubert 2009; Woszczyk et al. 2011), East Asia (Khim et al. 2005; Wang et al. 2012) and other regions (Dunn et al. 2008). Many factors may affect the TOC concentration in surface sediment, including productivity and sedimentation in the water column, inputs of terrestrial materials from the watershed, microbial activity and physicochemical properties of sediment (Burone et al. 2003; Gireeshkumar et al. 2013). Apparently, spatial variability of TOC is directly related to the contributions of various sources (i.e., autochthonous and allochthonous) that may vary largely between regions, partly owing to the diverse productivity and morphology (Anderson et al. 2009; Barnes and Barnes 1978).

There are a few approaches that have been employed to quantify organic carbon sources in sediments, including the isotope method ($\delta^{14}C$ and $\delta^{13}C$), element ratio method (C:N), and biomarkers (N-alkanes and fatty acids) (Bechtel and Schubert 2009; Fang et al. 2014; Hanson et al. 2014). One common approach is to apply a two (three)-end-member mixing model with the end-members for C:N ratio and $\delta^{13}C_{org}$ (stable carbon isotope in organic material). Such approach has been successfully used in studies of tracing biogeochemical process in various ecosystems (Liu and Kao 2007; Rumolo et al. 2011; Yu et al. 2010). Previous studies, based on such approach, have indicated that main sources of TOC are allochthonous in shallow and small lakes, but autochthonous in deep and large lakes (Barnes and Barnes 1978; Shanahan et al. 2013; Sifeddine et al. 2011).

Bosten Lake is the largest lake in Xinjiang, Northwest China, thus is a good place to study the lake carbon cycle. There have been limited studies conducted in Bosten Lake, which have focused evaluations on water quality (Wu et al. 2013), changes in lake level (Guo et al. 2014), carbon and oxygen isotopic composition of surface sediment carbonate (Zhang et al. 2009). In addition, a recent study has provided insights on the spatial and seasonal variations of particulate organic carbon (POC) for the water column of Bosten Lake (Wang et al. 2014), implying allochthonous contributions to sedimentary TOC. However, little is known about the magnitude and distribution of TOC, or the different contributions to TOC in the Bosten Lake. In addition to our previous analysis (Yu et al. 2015a), we conducted a further, comprehensive evaluation and discussion on the dynamics of TOC in surface sediment (0–2 cm) of Bosten Lake.

2 Sampling and Analyses

In August 2012, we used a Kajak gravity corer to collect surface (0–2 cm) sediments in the Bosten Lake's main section. There were 13 sampling locations that covered major parts of the lake, with the water depths ranging from 3 to 14 m (Fig. 1). Each

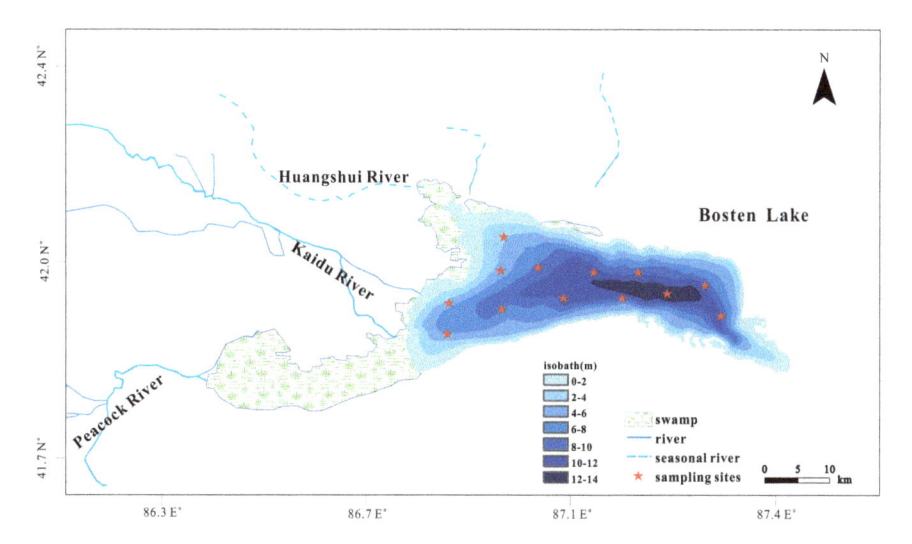

Fig. 1 Sampling locations and the water depth in the Bosten Lake. Bathymetric data was obtained from Wu et al. (2013) and the contours were plotted using software ArcGIS 10.1 and Corel DRAW X7

sediment core was sliced into 1-cm, then placed in polyethylene bags, and stored in a cooler until analyses.

Each sample (~0.5 g) was pretreated, with 10–20 ml of 30% H_2O_2 to remove organic matter in a water bath (60–80 °C), and then using 10–15 ml of 10% HCl to remove carbonates. The processed samples were mixed with 2000 ml of deionized water, and centrifuged after 24 h of standing. The solids were dispersed with 10–15 ml of 0.05 M $(NaPO_3)_6$, and then analyzed by a Malvern Mastersizer 2000 laser grain size analyzer. Related system software can automatically output the median diameter d(0.5), the diameter at the 50th percentile of the distribution (μm), and the percentages of clay (<2 μm), silt (2–64 μm) and sand (>64 μm) fractions, respectively.

Sediment TC and TN were measured by using an Elemental Analyzer 3000 (Euro Vector, Italy). Each sample was freeze-dried and ground into a fine powder, placed in tin capsules, then weighed and packed carefully. For the analysis of TOC, each sample (0.3–0.5 g) was pretreated with 5–10 ml 2 M HCl for 24 h at room temperature to remove carbonate, dried overnight under 60 °C, and analyzed using the elemental analyzer.

For the analyses of $\delta^{13}C_{org}$, we first removed carbonate with 5–10 ml 2 M HCl for 24 h at room temperature for all the freeze-dried sediment sample. After that, each pretreated sample was rinsed to a pH of ~7 with deionized water and dried at 40–60 °C. The samples were then combusted in the Thermo Elemental Analyzer integrated with an Isotope Ratio Mass Spectrometer (Delta Plus XP, Thermo Finnigan MAT, Germany). Isotopic data were reported in delta notation relative to the Vienna Pee Dee Belemnite (VPDB), with analytical precision of 0.1‰ for $\delta^{13}C_{org}$.

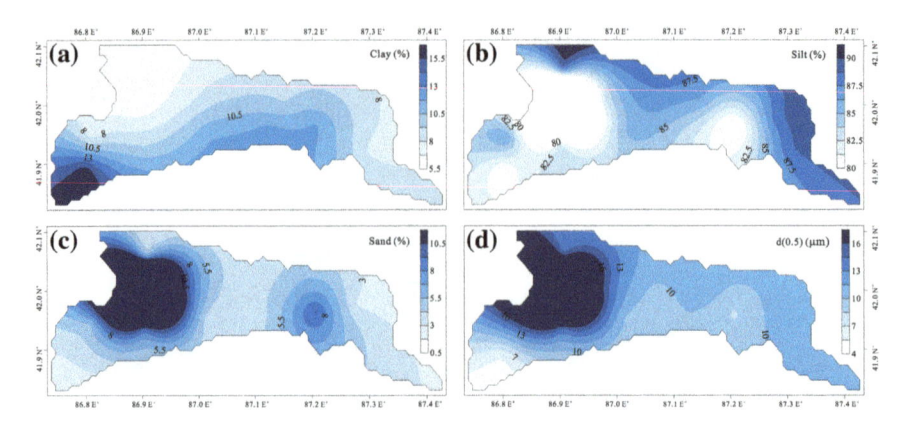

Fig. 2 Spatial variations of **a** clay, **b** silt, **c** sand and **d** the median diameter (d(0.5), μm) in the surface sediments of Bosten Lake. The distribution maps (Figs. 2–6) were produced using Surfer 9.0 (Golden Software Inc.) and the interpolated data in the maps were made by using the Krige Method of Gridding

3 Grain Size Characteristics

As shown in Fig. 2, there were large differences in the spatial distributions of the grain size in the surface sediment. Apparently, clay content (5.9–16.3%) was low in the whole lake, with the lowest to the north and relatively higher contents in the southern area. Much high value of silt content (>80%) was found in the sediments, with the highest near the mouths of the Kaidu River and Huangshui River, but the lowest silt content in the mid-west (between the rivers' mouths). On the other hand, the proportion of sand particles was relatively low with the highest content between the two rivers' mouths (Fig. 2c). As expected, similar to that of sand, the d(0.5) showed a spatial distribution with the highest values in the mid-west part, indicating much strong hydrodynamic effect in this section.

4 Spatial Distributions of TOC, TN, C:N and $\delta^{13}C_{org}$

There was a large range in the surface TOC, in which higher values (4.1–4.4%) were found in three parts, i.e., the northern lake section, the eastern lake section and the southwestern lake section (near the Kaidu River's mouth), and lower TOC values (2.1–2.9%) were in the mid-west lake section (Fig. 3a). Similarly, TN ranged from 0.28 to 0.62%, showing the highest in the northwest and east areas, and the lowest value in the mid-west lake area (Fig. 3b). Overall, TN exhibited similar spatial distribution to TOC, except in the northwest lake section where showed lower TOC value, but higher TN concentration.

Spatial Distribution of Organic Carbon in Surface Sediment ...

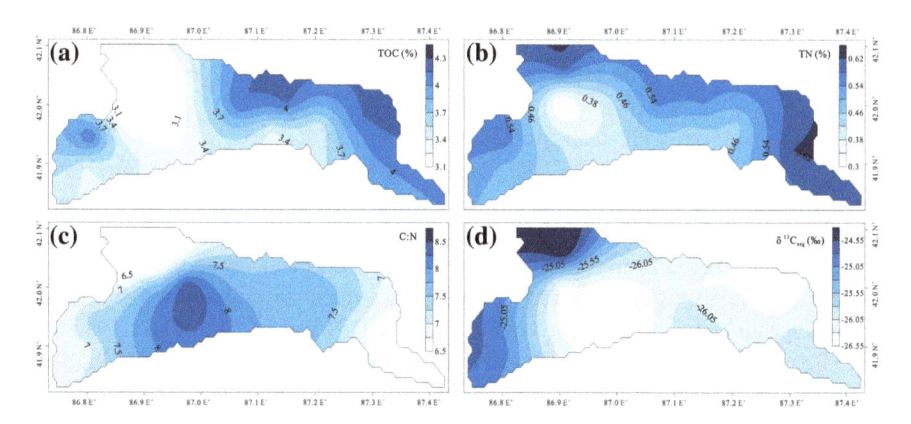

Fig. 3 Spatial distributions of **a** total organic carbon (TOC), **b** total nitrogen (TN), **c** C:N ratio (TOC:TN) and **d** carbon stable isotope ($\delta^{13}C_{org}$) of TOC in the surface sediments

As shown in Fig. 3c, C:N ratio (ranging from 4.8 to 8.5) revealed a large spatial variation in the surface sediment). Relative to other lake parts, the central part had generally higher C:N ratio, with the highest ratio found in the mid-west section, but the lowest near the mouth of Huangshui River. Value of $\delta^{13}C_{org}$ varied from −26.65 to −24.13‰ in the surface sediments, which was the least negative in the northwest lake part (close to the Huangshui River's mouth), but the most negative in the area between 86.9–87°E and 41.9–42°N. Overall, $\delta^{13}C_{org}$ value was much more negative in the central and eastern lake sections than in the western area in Bosten Lake.

5 Contributions of Different Sources

In this study, we used a three-end-member mixing model to quantify the contribution (f) of different sources (i.e., lake plankton, soil and high plant, denoted by 1, 2 and 3, respectively):

$$\delta = f_1\delta_1 + f_2\delta_2 + f_3\delta_3 \quad (1)$$

$$r = f_1 r_1 + f_2 r_2 + f_3 r_3 \quad (2)$$

$$1 = f_1 + f_2 + f_3 \quad (3)$$

where δ and r were $\delta^{13}C_{org}$ value and C:N ratio, respectively.

Firstly, we assumed that the main native plants were responsible for high plant's contribution, because there were limited crops growing around the lake and crops' growing season was short (most was less than five months). According to the recent survey in Yanqi Basin (Zhang 2013), the mean C:N ratio was 10.0 and 22.1, and

$\delta^{13}C_{org}$ value was −23.6 and −26.4‰ for the surface soil and native plant around the lake, respectively. Thus, we used these values as our end members in the mixing model.

In addition, there were measurements of POC and its $\delta^{13}C_{org}$, and PON (particulate organic nitrogen) in the lake water column (Wang et al. 2014). The mean concentrations of POC and PON showed an increase, from 0.61 mg C L^{-1} and 0.072 mg N L^{-1} in spring to 0.70 mg C L^{-1} and 0.088 mg N L^{-1} in summer, and $\delta^{13}C_{org}$ value changed from −22.9‰ in spring to −23.5‰ in summer. We assumed that lake plankton productivity was responsible for the changes in these parameters between spring and summer and estimated that the C:N and $\delta^{13}C_{org}$ of lake plankton was 5.3 and −27.7‰, which were used as the end-members in the calculation.

Figure 4 showed the contributions of autochthonous and allochthonous sources to TOC in the surface sediment. Lake plankton's contribution ranged from 54 to 90%, with the smallest contribution found in the southern and eastern deep sections, but the largest contribution in the western shallow section. Soil's contribution varied from 10 to 40%, and the largest contribution was observed in the southeastern lake area and near central-south lake area, but the smallest in the southwestern area. Clearly, native plant had extremely low contribution to TOC, with only a few sites having contributions greater 10–12% in Fig. 4c. The average contributions of lake plankton, soil and native plant to TOC were 66, 30 and 4%, respectively.

Figure 5 revealed significant differences in the spatial pattern between the autochthonous TOC and allochthonous TOC. The autochthonous TOC showed both the highest (~3.5%) and lowest value (~1.5%) in the western section, with the highest found near the river inlet (Fig. 5a). On the other hand, autochthonous sources presented a clear decrease trend from north to south in the area east of 87°E. Overall, the allochthonous TOC showed an apparent elevation from 0.5% in the west to 1.9% in the east (Fig. 5b).

6 Variation of TOC and Relationship with Water Column POC

Our analyses demonstrated that the highest autochthonous TOC was found near the Kaidu River's mouth, but the highest allochthonous TOC in the eastern part of the lake (Fig. 5), which might be associated with the spatial distributions of nutrients in Bosten Lake. The transportation of various materials through the Kaidu River would have a significant downward trend from the west river mouth to the east lake section, which could largely influence the nutrient conditions in Bosten Lake. In general, lake productivity was high near the sources of nutrients, such as estuaries owing to the extra input of riverine nutrient (Deng et al. 2006; Lin et al. 2002).

To better understand the dynamics of TOC burial in Bosten Lake, we evaluated the spatial distribution of water column POC in summer and fall (Fig. 6). There were large differences in the magnitude and spatial pattern of POC between the

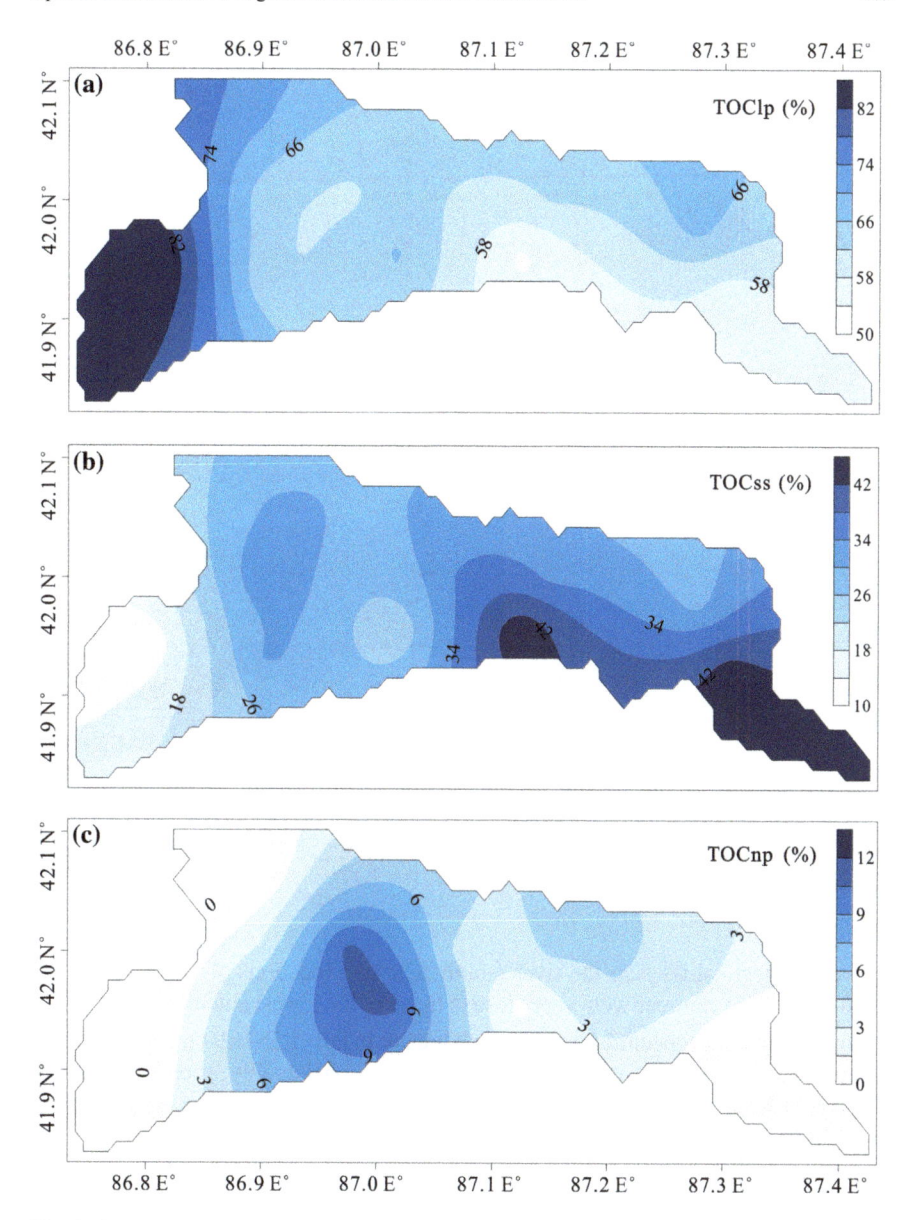

Fig. 4 Spatial variabilities of relative contributions (percentages) to the TOC in surface sediment of Bosten Lake. **a** TOC from the lake plankton (TOC$_{lp}$), **b** TOC from the surface soil (TOC$_{ss}$), and **c** TOC from the native plant (TOC$_{np}$)

two seasons. Overall, there was a similarity in the spatial pattern between summer POC and autochthonous TOC (see Fig. 5a) in spite of differences. Both variables

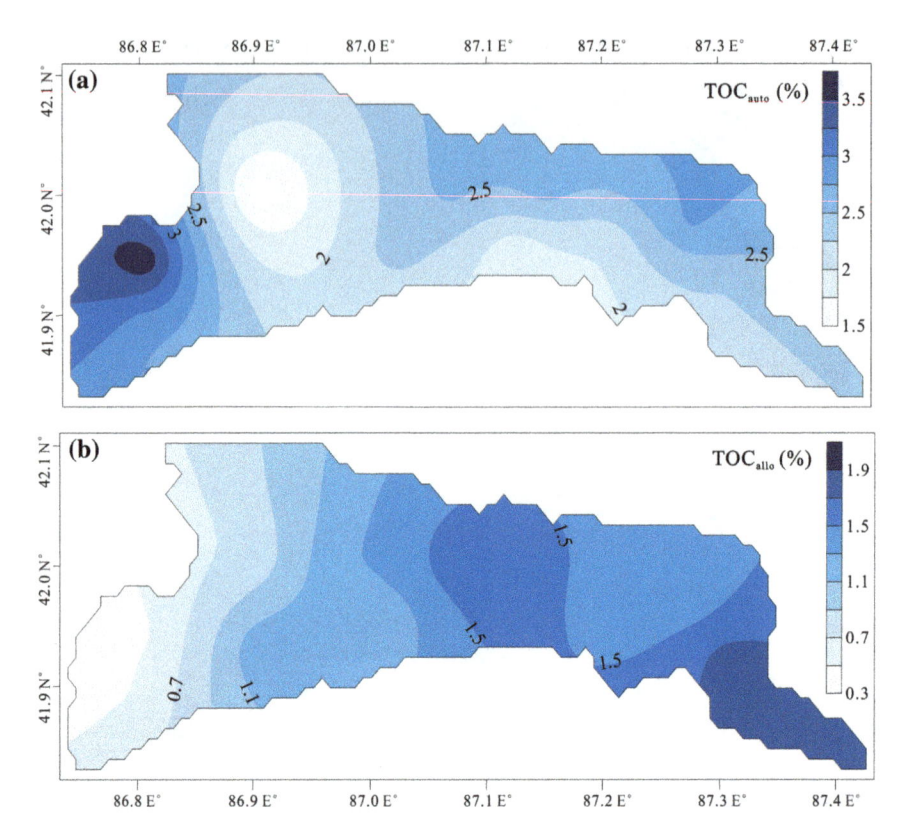

Fig. 5 Spatial patterns of **a** autochthonous sources TOC (TOC$_{auto}$) and **b** allochthonous sources TOC (TOC$_{allo}$) in surface sediment

showed highest values near the river mouth. The two locations of the lowest POC (<0.6 mg L^{-1}) in summer were very close to those of the lowest autochthonous TOC.

Given that POC concentration was much greater in fall (1.35–1.94 mg L^{-1}) than in summer (0.4–1.03 mg L^{-1}), one might assume that a considerable proportion of POC in fall was from allochthonous sources. Interestingly, there were large differences in the spatial pattern between fall POC and allochthonous TOC (see Fig. 5b). For example, fall POC revealed highest value in the central part and lowest value in the far east of the lake whereas allochthonous TOC showed highest values in the eastern section; and there was a north-to-south elevation in the former but a west-to-east elevation in the latter. Assuming that the level of autochthonous POC was the same in summer and in fall, one might take the difference of POC (ΔPOC) between fall and summer as allochthonous. However, the spatial variation of ΔPOC was different from that of allochthonous TOC, but similar to that of fall POC. While elevated POC in fall might be a result of allochthonous contribution, the spatial distribution of TOC burial could be affected by various processes (see discussion in next section).

Spatial Distribution of Organic Carbon in Surface Sediment ... 111

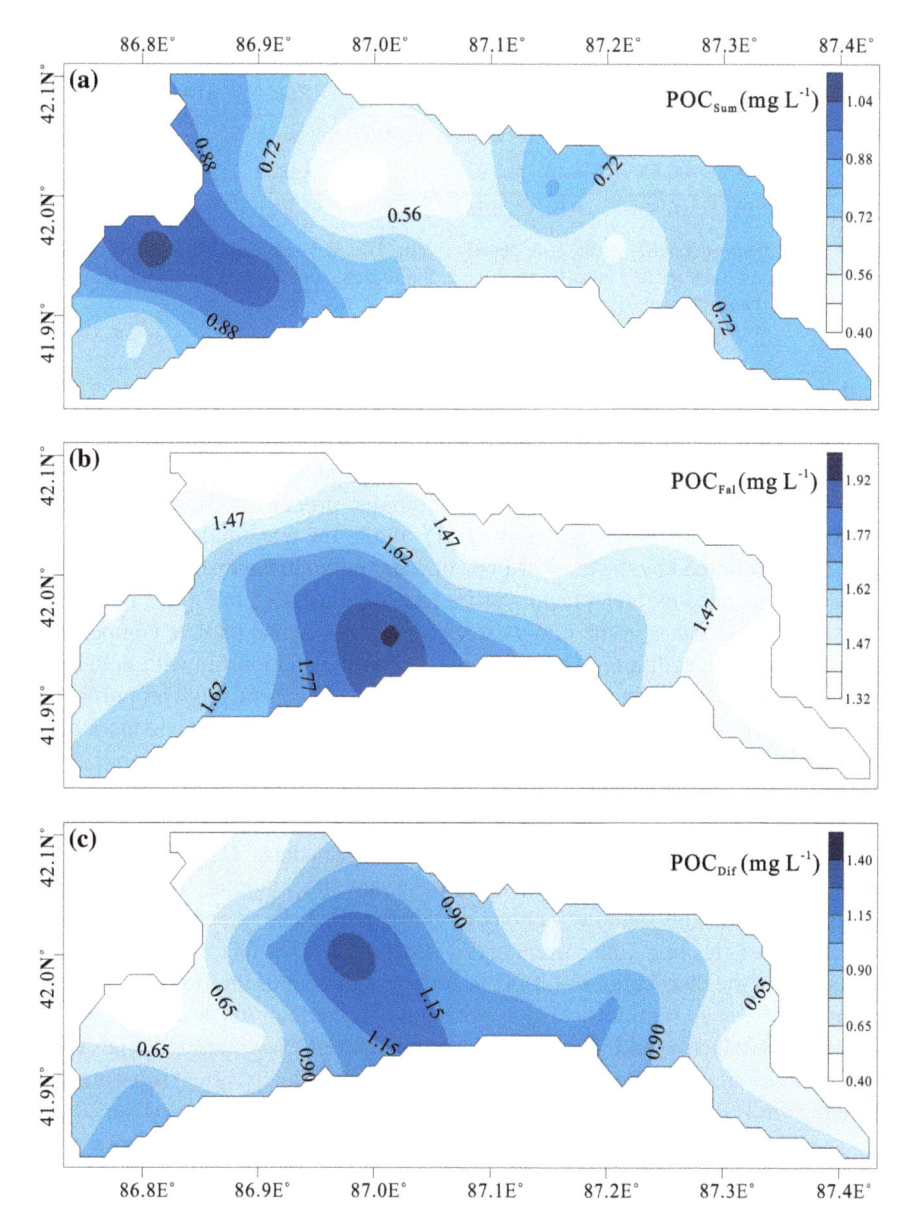

Fig. 6 Spatial distribution of POC concentration in the water column of Bosten Lake in **a** summer, **b** fall and **c** the difference. Data were from Wang et al. (2014)

Table 1 Correlation coefficient (r) between variables in surface sediment

Variables	WD	DBD	d(0.5)	Clay	Silt	Sand	TOC	$\delta^{13}C_{org}$
TOC	0.50	−0.58*	−0.71**	0.18	0.77**	−0.76**		−0.15
$\delta^{13}C_{org}$	−0.66*	−0.46	−0.13	0.03	0.21	−0.20	−0.15	
C:N	0.50	0.50	0.01	0.25	−0.19	0.11	0.14	−0.82**

WD = water depth (m), DBD = dry bulk density (g cm^{-3}), d(0.5) = median diameter (μm) and clay, silt and sand fractions (%) in surface sediment. Significance of Pearson Correlation is marked with *(p <0.05) and **(p <0.01) superscripts

7 Dynamics of TOC and Underlying Mechanisms

Previous studies have indicated that the magnitudes and spatial distribution of TOC in lake sediment may reflect multiple and complex processes (Dunn et al. 2008; Sifeddine et al. 2011; Woszczyk et al. 2011). While the magnitude of TOC burial was often determined by the level of POC in water column, the distribution of TOC might be influenced by many factors/processes. As illustrated in Figs. 2 and 3, the lowest TOC was found in the mid-west section where coarse particle components were dominant. In addition, TOC expressed a negative relationship with both sand content and d(0.5) value in surface sediment (Table 1). In general, high proportion of coarser particles reflected an environment with stronger water energy and more finer particles indicated stable sedimentary conditions (Jin et al. 2006; Molinaroli et al. 2009). Thus, it appeared that lower TOC concentration in the mid-west lake part was induced by both the lower level of POC and higher hydrodynamic circumstances.

Based on the significant negative relationship between the $\delta^{13}C_{org}$ value and water depth (Table 1), we could imply that there was more allochthonous TOC (with less negative $\delta^{13}C$) in the shallow sections of the Bosten Lake. Apart from the lake own characteristics (e.g., the current and water depth), other factors may also have influences on the distributions of TOC in surface sediment of the Bosten Lake. For example, there were significant land use changes (i.e., agricultural development with fertilization), which would increase the riverine input of exogenous materials (particularly nutrients), leading to variations in lake water productivity and subsequently changing the TOC storage in surface sediment (Lami et al. 2010; Rumolo et al. 2011).

8 Implications and Future Directions

There are obvious differences in TOC values of surface sediment across lake regions in China. For example, lower concentration (0.2–2%) was observed in lakes in the Tibetan Plateau (Lami et al. 2010; Wang et al. 2012) and in the Yangtze floodplain (Dong et al. 2012; Wu et al. 2007), but much higher concentration (5–13%) was found in the Yunnan-Guizhou Plateau (Wu et al. 2012; Zhu et al. 2013). The TOC content is modestly high in Bosten Lake (1.8–4.4%), comparing with the lakes across China.

Climate change and human activities in the Yanqi Basion have resulted in remarkable changes in many aspects, including changes in the runoff of the Kaidu River and the lake level of Bosten Lake over the past decades (Guo et al. 2014), which would have impacted the spatial and temporal variations of TOC burial in the sediment of Bosten Lake. A recent study demonstrates that there has been pronounced interannual variability in carbon burial not only for TOC but also carbonate, particularly post 2000 (Yu et al. 2015b). Further studies are needed to evaluate the spatial-temporal variations of the water column biological production and carbon burial in sediments to better understand the dynamics of organic and inorganic carbon in different lake regions and the impacts of human disturbance and climate change.

9 Conclusion

Based on the surface (0–2 cm) sediment samples, we found a large spatial variability in TOC content (2.1–4.2%) and $\delta^{13}C_{org}$ value (-26.65 to -24.13‰) in the Bosten Lake. Using a three end member mixing model, we estimated that 54–90% of TOC was from autochthonous sources. Higher TOC concentration (>3.7%) was found near the area of the Kaidu River mouth, in the central-north section and in the east section which was attributable to autochthonous, autochthonous plus allochthonous, and allochthonous sources, respectively. The lowest TOC content was found in the mid-west section, which might mainly be a result of high kinetic energy levels. Our study indicated that multiple and complex processes affected the TOC dynamics in surface sediment of the Bosten Lake.

References

Anderson NJ, D'Andrea W, Fritz SC (2009) Holocene carbon burial by lakes in SW Greenland. Glob Change Biol 15:2590–2598. https://doi.org/10.1111/j.1365-2486.2009.01942.x

Barnes MA, Barnes WC (1978) Organic compounds in lake sediments. In: Lerman A (ed) Lakes. Springer, New York

Battin TJ, Luyssaert S, Kaplan LA, Aufdenkampe AK, Richter A, Tranvik LJ (2009) The boundless carbon cycle. Nat Geosci 2:598–600

Bechtel A, Schubert CJ (2009) A biogeochemical study of sediments from the eutrophic Lake Lugano and the oligotrophic Lake Brienz, Switzerland. Org Geochem 40:1100–1114

Burone L, Muniz P, Pires-Vanin A, Maria S, Rodrigues M (2003) Spatial distribution of organic matter in the surface sediments of Ubatuba Bay (Southeastern-Brazil). An Acad Bras Ciênc 75:77–80

Dean WE, Gorham E (1998) Magnitude and significance of carbon burial in lakes, reservoirs, and peatlands. Geology 26:535–538

Deng B, Zhang J, Wu Y (2006) Recent sediment accumulation and carbon burial in the East China Sea. Glob Biogeochem Cycles 20:GB3014

Dong X, Anderson NJ, Yang X, Shen J (2012) Carbon burial by shallow lakes on the Yangtze floodplain and its relevance to regional carbon sequestration. Glob Change Biol 18:2205–2217

Downing JA, Prairie Y, Cole J, Duarte C, Tranvik L, Striegl R, McDowell W, Kortelainen P, Caraco N, Melack J (2006) The global abundance and size distribution of lakes, ponds, and impoundments. Limnol Oceanogr 51:2388–2397

Dunn RJK, Welsh DT, Teasdale PR, Lee SY, Lemckert CJ, Meziane T (2008) Investigating the distribution and sources of organic matter in surface sediment of Coombabah Lake (Australia) using elemental, isotopic and fatty acid biomarkers. Cont Shelf Res 28:2535–2549

Fang J, Wu F, Xiong Y, Li F, Du X, An D, Wang L (2014) Source characterization of sedimentary organic matter using molecular and stable carbon isotopic composition of n-alkanes and fatty acids in sediment core from Lake Dianchi, China. Sci Total Environ 473:410–421. https://doi.org/10.1016/j.scitotenv.2013.10.066

Gireeshkumar TR, Deepulal PM, Chandramohanakumar N (2013) Distribution and sources of sedimentary organic matter in a tropical estuary, south west coast of India (Cochin estuary): a baseline study. Mar Pollut Bull 66:239–245

Guo M, Wu W, Zhou X, Chen Y, Li J (2014) Investigation of the dramatic changes in lake level of the Bosten Lake in northwestern China. Theor Appl Climatol 1–11

Hanson PC, Buffam I, Rusak JA, Stanley EH, Watras C (2014) Quantifying lake allochthonous organic carbon budgets using a simple equilibrium model. Limnol Oceanogr 59:167–181

Jin Z, Li F, Cao J, Wang S, Yu J (2006) Geochemistry of Daihai Lake sediments, Inner Mongolia, north China: implications for provenance, sedimentary sorting, and catchment weathering. Geomorphology 80:147–163

Khim B-K, Jung HM, Cheong D (2005) Recent variations in sediment organic carbon content in Lake Soyang (Korea). Limnology 6:139

Lami A, Turner S, Musazzi S, Gerli S, Guilizzoni P, Rose N, Yang H, Wu G, Yang R (2010) Sedimentary evidence for recent increases in production in Tibetan plateau lakes. Hydrobiologia 648:175–187

Lin S, Hsieh IJ, Huang K-M, Wang C-H (2002) Influence of the Yangtze River and grain size on the spatial variations of heavy metals and organic carbon in the East China Sea continental shelf sediments. Chem Geol 182:377–394

Liu K-K, Kao S-J (2007) A three end-member mixing model based on isotopic composition and elemental ratio. Terr Atmos Oceanic Sci 18

Molinaroli E, Guerzoni S, De Falco G, Sarretta A, Cucco A, Como S, Simeone S, Perilli A, Magni P (2009) Relationships between hydrodynamic parameters and grain size in two contrasting transitional environments: The Lagoons of Venice and Cabras, Italy. Sed Geol 219:196–207

Rumolo P, Barra M, Gherardi S, Marsella E, Sprovieri M (2011) Stable isotopes and C/N ratios in marine sediments as a tool for discriminating anthropogenic impact. J Environ Monit 13:3399–3408

Shanahan TM, McKay N, Overpeck JT, Peck JA, Scholz C, Heil CW Jr, King J (2013) Spatial and temporal variability in sedimentological and geochemical properties of sediments from an anoxic crater lake in West Africa: implications for paleoenvironmental reconstructions. Palaeogeogr Palaeoclimatol Palaeoecol 374:96–109. https://doi.org/10.1016/j.palaeo.2013.01.008

Sifeddine A, Meyers P, Cordeiro R, Albuquerque A, Bernardes M, Turcq B, Abrão J (2011) Delivery and deposition of organic matter in surface sediments of Lagoa do Caçó (Brazil). J Paleolimnol 45:385–396

Tranvik LJ, Downing JA, Cotner JB, Loiselle SA, Striegl RG, Ballatore TJ, Dillon P, Finlay K, Fortino K, Knoll LB (2009) Lakes and reservoirs as regulators of carbon cycling and climate. Limnol Oceanogr 54:2298–2314

Wang Y, Zhu L, Wang J, Ju J, Lin X (2012) The spatial distribution and sedimentary processes of organic matter in surface sediments of Nam Co, Central Tibetan Plateau. Chin Sci Bull 57:4753–4764

Wang X, Fang C, Yu Z, Wang J, Peng D, Jingjing K (2014) Seasonal variations of particulate and dissolved organic carbon in Bosten Lake, Xinjiang. J Lake Sci 26:552–558

Woszczyk M, Bechtel A, Gratzer R, Kotarba MJ, Kokociński M, Fiebig J, Cieśliński R (2011) Composition and origin of organic matter in surface sediments of Lake Sarbsko: a highly eutrophic and shallow coastal lake (northern Poland). Org Geochem 42:1025–1038. https://doi.org/10.1016/j.orggeochem.2011.07.002

Wu J, Huang C, Zeng H, GerhardH S, Rick B (2007) Sedimentary evidence for recent eutrophication in the northern basin of Lake Taihu, China: human impacts on a large shallow lake. J Paleolimnol 38:13–23

Wu F, Xu L, Sun Y, Liao H, Zhao X, Guo J (2012) Exploring the relationship between polycyclic aromatic hydrocarbons and sedimentary organic carbon in three Chinese lakes. J Soils Sediments 12:774–783

Wu J, Ma L, Zeng H (2013) Water quality and quantity characteristics and its evolution in Lake Bosten, Xinjiang over the past 50 years. Sci Geogr Sinica 33:231–237

Yu F, Zong Y, Lloyd JM, Huang G, Leng MJ, Kendrick C, Lamb AL, Yim WWS (2010) Bulk organic $\delta 13C$ and C/N as indicators for sediment sources in the Pearl River delta and estuary, southern China. Estuar Coast Shelf Sci 87:618–630

Yu ZT, Wang XJ, Zhang EL, Zhao CY, Liu XQ (2015a) Spatial distribution and sources of organic carbon in the surface sediment of Bosten Lake, China. Biogeosciences 12:6605–6615. https://doi.org/10.5194/bg-12-6605-2015

Yu ZT, Wang XJ, Zhao CY, Lan HY (2015b) Carbon burial in Bosten Lake over the past century: impacts of climate change and human activity. Chem Geol 419:132–141. https://doi.org/10.1016/j.chemgeo.2015.10.037

Zhang J (2013) Impacts of land use on soil organic matter in Yanqi Basin. Xinjiang Institute of Ecology and Geography, Chinese Academy of Sciences, Urumqi, Xinjiang, P.R. China

Zhang C, Mischke S, Zheng M, Prokopenk A, Guo F, Feng Z (2009) Carbon and oxygen isotopic composition of surface-sediment carbonate in Bosten Lake (Xinjiang, China) and its controlling factors. Acta Geol Sinica-English Edition 83:386–395. https://doi.org/10.1111/j.1755-6724.2009.00029.x

Zhu Z, Ja Chen, Zeng Y (2013) Abnormal positive delta C-13 values of carbonate in Lake Caohai, southwest China, and their possible relation to lower temperature. Quatern Int 286:85–93

Temporal Variability of Carbon Burial and the Underlying Mechanisms in Bosten Lake Since 1950

Zhitong Yu and Xiujun Wang

Abstract The arid and semi-arid regions of northwest China have experienced significant climate changes and human activities since 1950, with implications for the carbon cycle. We collected two short sediment cores to investigate the anthropogenic and climate impacts on the carbon burial in Bosten Lake. Basic physical and chemical properties were measured, including total organic carbon (TOC) and total inorganic carbon (TIC), isotopic compositions of TOC ($\delta^{13}C_{org}$) and carbonate ($\delta^{13}C_{carb}$ and $\delta^{18}O_{carb}$), grain size, and ^{137}Cs and ^{210}Pb. Our data showed carbon burials revealed a profound temporal variability with an overall increasing trend, and the TIC burial rate was proximately twice of the TOC since 1950. The remarked increasing trend in TOC burial rate since 1950 (TIC burial during 1950–2002) was a result of warming and/or increased human activities; the sharp decline in TIC during the recent decade (2002–2012) was associated with lower biological activity, reduced evaporation and a rapid decline in lake level. The role of carbon storage in lake sediments, especially in arid area, should deserve to be paid more attention to the terrestrial carbon cycle.

1 Introduction

Lakes often receive a large amount of terrestrial materials from the watersheds, including organic carbon and nutrients. Organic carbon inputs can lead to more carbon burial in the sediments, whereas nutrient inputs can enhance biological activity in the water column and promote carbon burial through sedimentation (Cole et al. 2007). Earlier analyses showed that on the global scale, carbon burial in lake sediments ranged from 0.03 to 0.07 Pg C year^{-1} (Dean and Gorham 1998; Einsele

---

Z. Yu (✉) · X. Wang
College of Global Change and Earth System Science,
Beijing Normal University, Beijing 100875, China
e-mail: yuzhitong@bnu.edu.cn

Z. Yu
Xinjiang Institute of Ecology and Geography, Chinese Academy of Sciences,
Urumqi 830011, China

© Springer Nature Singapore Pte Ltd. 2018
X. Wang et al. (eds.), *Carbon Cycle in the Changing Arid Land of China*,
Springer Earth System Sciences, https://doi.org/10.1007/978-981-10-7022-8_9

et al. 2001). The amount of carbon burial in lakes is ~3% of the carbon sink by the global ocean, although the total coverage by lakes is only 0.8% of the total ocean area (Battin et al. 2009). Thus, lakes play an important role in the global carbon cycle and climate mitigation.

Sediment carbon consists of total organic carbon (TOC) and inorganic carbon (TIC). Many studies have been conducted to evaluate the rates of TOC in lake sediments over the twentieth century. Some studies showed large spatial variability in TOC burial (Anderson et al. 2013; Dong et al. 2012; Gui et al. 2013), and other analyses revealed an increasing trend in TOC burial rate during the past century (Anderson et al. 2014; Herczeg et al. 2001; O'Beirne et al. 2015). In particular, TOC burial rate was higher than 50 g C m^{-2} year^{-1} in the lakes of Europe in the last half-century. However, little has been done to assess the dynamics of TIC burial in lake sediments.

Early studies have demonstrated that variation of carbon burial in lakes is related to many factors, including lake area (Kastowski et al. 2011), sedimentary condition (Blais and Kalff 1995), watershed area (Downing et al. 2008), nutritional status (Anderson et al. 2014; Heathcote and Downing 2012), geographical location and human activity (Alin and Johnson 2007). Some analyses showed that temperature played a major role in controlling TOC burial in the Asia (Xu et al. 2013), whereas other analyses on lakes in the Europe and North America suggested that human activity was responsible for the increasing trend of TOC burial in lake sediments (Anderson et al. 2013, 2014). Further studies are needed to better understand the spatial and temporal variations of TOC burial under different impacts of climatic and anthropogenic factors. Moreover, there has been lacking data of quantifying both TOC and TIC burial rates.

The arid region of northwest China, particularly Xinjiang, has been undergone remarkable climate changes and human activities since 1950. For example, both temperature and precipitation presented a significant increasing trend over the second half of century in Xinjiang (Li et al. 2013), which has impacts on the hydrological and geochemical processes over various scales. In particular, increased precipitation and warming-induced glacier melting would lead to an increase in watershed's runoff (Wang et al. 2003), with implications for lakes in many aspects. On the other hand, there has been an increased water usage due to industrial and anthropogenic activities in Yanqi Basin and surrounding areas, which led to a significant drop in the lake level of the Bosten Lake (Guo et al. 2015). These changes can have large impacts on the carbon burial in Bosten Lake.

To investigate the impacts of climate change and human activities on the carbon burial in the Bosten Lake, we use the data from two short sediment cores (23.5 cm in length at B3 and 21.5 cm at B5) which were published in Chemical Geology (Yu et al. 2015a), with a focus on the period of 1950–2012. The site B5 is located at the deepest location, which is close to the BT04C site and BST16 site that were sampled in 2004 (Chen et al. 2006) and 2010 (Zheng 2012), respectively (Fig. 1). We also collect hydrological and climate variables for the Yanqi Basin and Bosten Lake. The objective of this study is to improve our understanding of the temporal variability of both TOC and TIC in response to the interactive effects of climate change and human

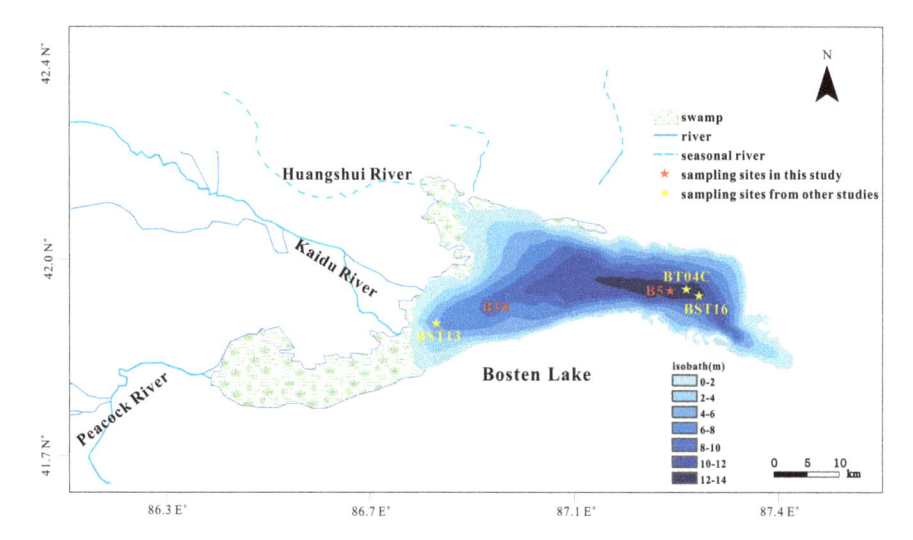

Fig. 1 Map of isobaths and sampling sites in the Bosten Lake. B3 and B5 from Yu et al. (2015a), BT04C from Chen et al. (2006) and BST 13 and BST16 from Zheng (2012). Bathymetric contours were plotted by ArcGIS 10.0 and CorelDRAW X7

activities post 1950, by examining the chronology and multi-proxies (i.e. grain size, organic carbon, inorganic carbon and stable carbon isotopes) of lacustrine sediment in Bosten Lake.

2 Chronology and Sedimentation

Following the standard procedures, sediment samples were analysed for ^{210}Pb, ^{226}Ra and ^{137}Cs by using Ortec HPGe GWL series well-type coaxial low background intrinsic germanium detectors. ^{210}Pb was measured by its gamma emissions at 46.5 keV, and ^{226}Ra by 295 and 352 keV γ-rays emitted by its daughter isotope ^{214}Pb following 3 weeks storage in sealed containers in order to allow radioactive equilibration. ^{137}Cs was determined via its emissions at 662 keV (Appleby 2001).

Figure 2 shows the profiles of original ^{137}Cs activities at the newer sites (B3 and B5), BT04C from Chen et al. (2006), and the BST13 and BST16 sites from Zheng (2012). The ^{137}Cs curve showed several peaks at both B3 and B5 sites, which was also seen at other sites in Bosten Lake. As it was known that two peaks in the ^{137}Cs curve were widely recognized, which resulted from the global fallout peaks in 1963 and the Chernobyl accident in 1986 (Appleby 2001). Earlier studies pointed out that the artificial radionuclide in the Bosten Lake was not only influenced by the global weapon testing, but also by the regional weapon testing near the study site (Chen et al. 2006). From this point of view, we dated the deepest obvious peak of ^{137}Cs as the year 1963, which was at 18.5 and 17 cm for the B3 site and B5 site, respectively.

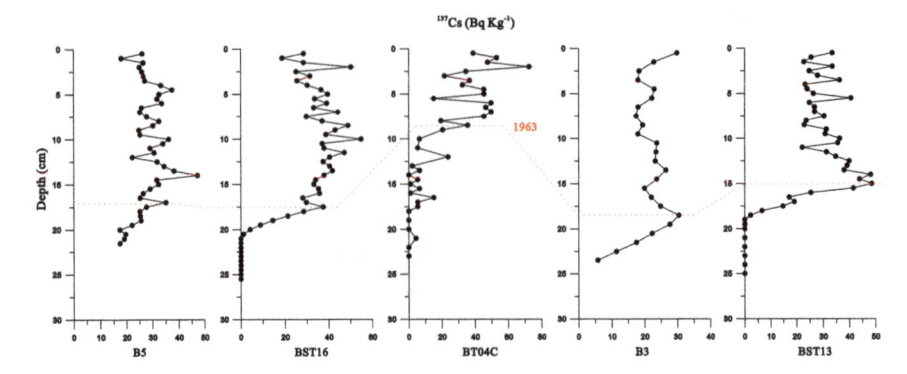

Fig. 2 Changing sequence of ^{137}Cs over depth in different cores in the Bosten Lake

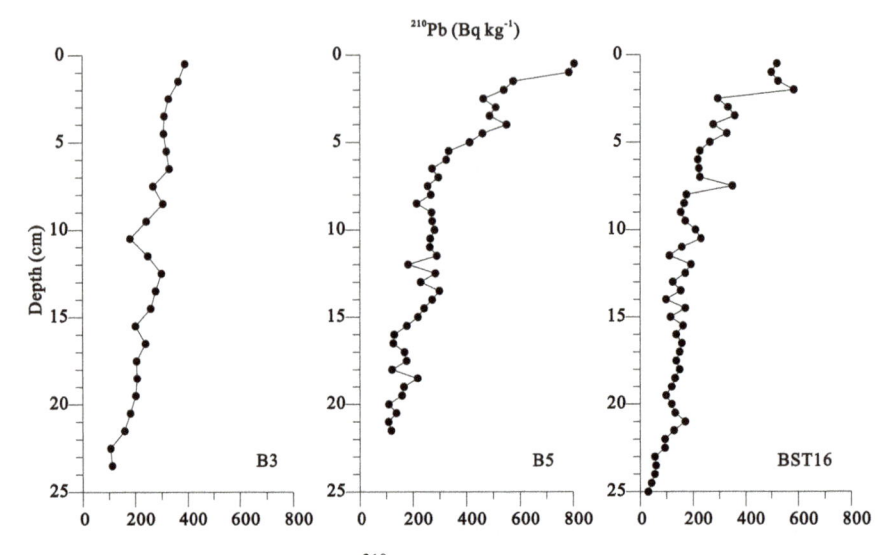

Fig. 3 Vertical distributions of excess ^{210}Pb in the Bosten Lake

It was worth noting that we considered the next obvious peak of ^{137}Cs (with the highest activity at around 14 cm at the B5 site) corresponded the year 1976, because there were the most intensive nuclear tests in China carried out in 1976, according to the local weapon testing records (Chen et al. 2006).

As shown in Fig. 3, excess ^{210}Pb revealed an exponential decrease with depth, which was similar to the profiles observed in the previous studies (Chen et al. 2006; Zheng 2012) carried out in the Bosten Lake.

We calculated the dating using the composite model (Appleby 2001). Firstly, we used the peak of ^{137}Cs near the base of the core to determine the depth corresponding to the year 1963, and then, we determined the year Y_Z of each sediment layer between the surface and the 1963-depth according to the following equations:

$$Y_z = Y_0 + \frac{1}{\lambda} \ln\left(1 + \frac{\lambda(A_0 - A_z)}{P}\right) \quad (1)$$

$$P = \frac{-\lambda(A_0 - A_w)}{1 - e^{\lambda(T_0 - 1963)}} \quad (2)$$

where Y_0 was the sampling year 2012, and λ the decay constant of ^{210}Pb (i.e. 0.03114 year^{-1}). The terms A_w, A_0 and A_z were the inventory of excess ^{210}Pb for the section below the 1963-depth, the entire core and the section below depth z, respectively.

The dates for the layers below the 1963-depth were calculated based on:

$$Y_z = 1963 - \frac{1}{\lambda} \ln\left(\frac{A_w}{A_z}\right) \quad (3)$$

The sedimentation rate (SR, g cm^{-2} year^{-1}) was determined as:

$$SR = \frac{d\Delta z}{\Delta T} \quad (4)$$

where d was the dry bulk density (g cm^{-3}).

Using the composite model, we dated the B3 and B5 cores with a time span of ~100 and ~110 years, respectively. To evaluate the dating results, we examined the relationship between the temporal variability of $\delta^{18}O_{carb}$ from the B5 site and the time series of lake evaporation, since higher rate of evaporation would cause ^{18}O enriched in the lake water and thus in sediment (Zhang et al. 2009). We found that the value of $\delta^{18}O_{carb}$ showed a good relationship with evaporation ($r = 0.69$, $P < 0.001$), indicating that our chronology was reasonable.

There was a generally increasing trend in the sedimentation rate at both sites over the past century, with a pronounced increase post 1950 (Fig. 4). The mean sedimentation rate was 0.14 and 0.10 g m^{-2} year^{-1} at the site B5 and B3, respectively. In addition to the general increasing trend at the B5 site, there were a few peaks in sedimentation, i.e. 0.13, 0.19 and 0.24 g m^{-2} year^{-1} around 1970, 1985 and 2000.

3 Vertical Variations of the Sedimentary Properties

Dry bulk density (DBD) showed an increasing trend with depth, with very low values (0.1–0.2 g cm^{-3}) in the surface and much higher values (0.4–0.9 g cm^{-3}) near the bottom (Fig. 5). In general, DBD was higher at the B5 site than at the B3 site; there was little change in the upper section at the B3 site, but a sharp increase with depth in the upper 7 cm at the B5 site. The median diameter ($d_{0.5}$) showed obvious differences (Tables 1 and 2), with much higher values at the B3 site, indicating that the kinetic energy was stronger at the B3 site, perhaps due to the inflow and outflow near the river mouths. Generally, there was a lack of clear vertical variation of $d_{0.5}$ value at

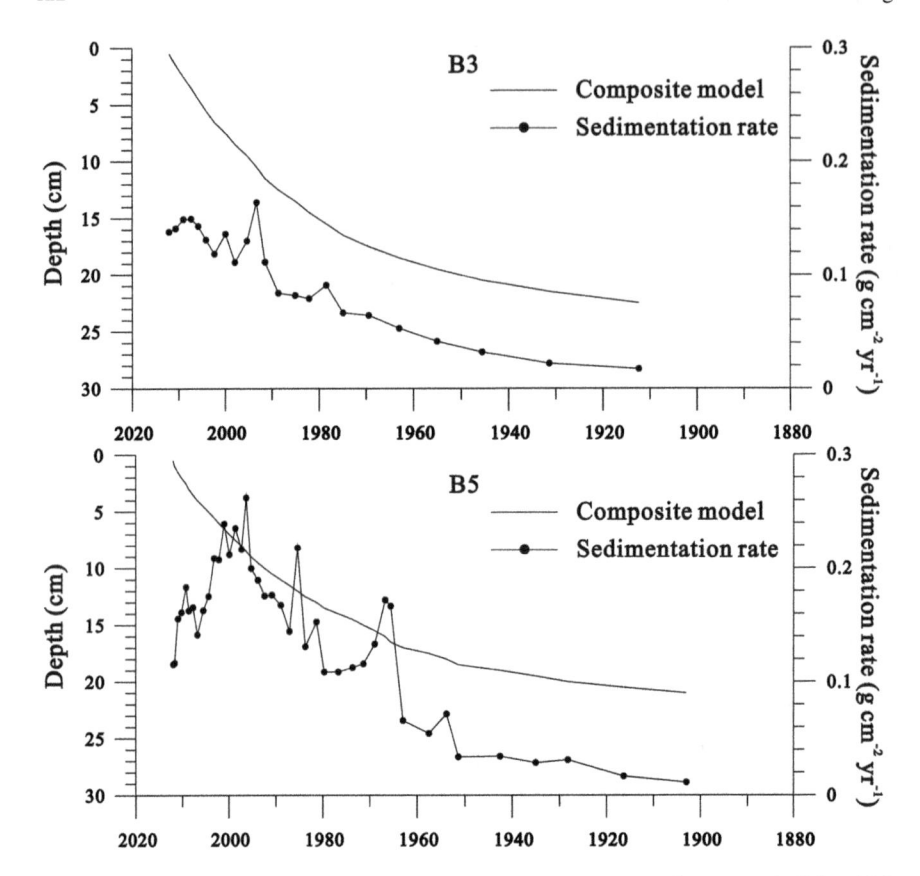

Fig. 4 Correspondence between depth and year and dry mass accumulation rate at the B3 and B5 sites

both B3 and B5 sites, particularly above 20 cm, which indicated that there might have been little changes in the hydrodynamic processes over the past decades.

There was a decreasing trend with depth in TOC at both sites, and the decrease was exponential at the B5 site, but almost linear at the B3 site (Fig. 5). On average, TOC content was significantly lower at the B5 site (2.0 ± 0.6%) than that at the B3 site (2.9 ± 0.4%) (Tables 1 and 2). Particularly, the difference was remarkable in the subsurface, i.e. <1.8% at the B5 site, but 2.3–3.1% at the B3 site. $\delta^{13}C_{org}$ revealed an opposite vertical distribution at the two sites that had much different $\delta^{13}C_{org}$ values in the upper 6 cm (i.e. −27.9‰ at the B3 and −26.5‰ at the B5 site), but similar values (i.e. −27.4‰) below 19 cm.

As shown in Fig. 5, the magnitude and vertical distribution of TIC were somehow different between the two sites. Clearly, TIC content was relatively higher in the upper section at the B3 site (5.5 ± 0.3%), but in the lower section at the B5 site (5.9 ± 0.1%) (Tables 1 and 2). Overall, there was an increasing trend with depth at the B5 site, but little vertical variation at the B3 site. $\delta^{13}C_{carb}$ revealed a generally

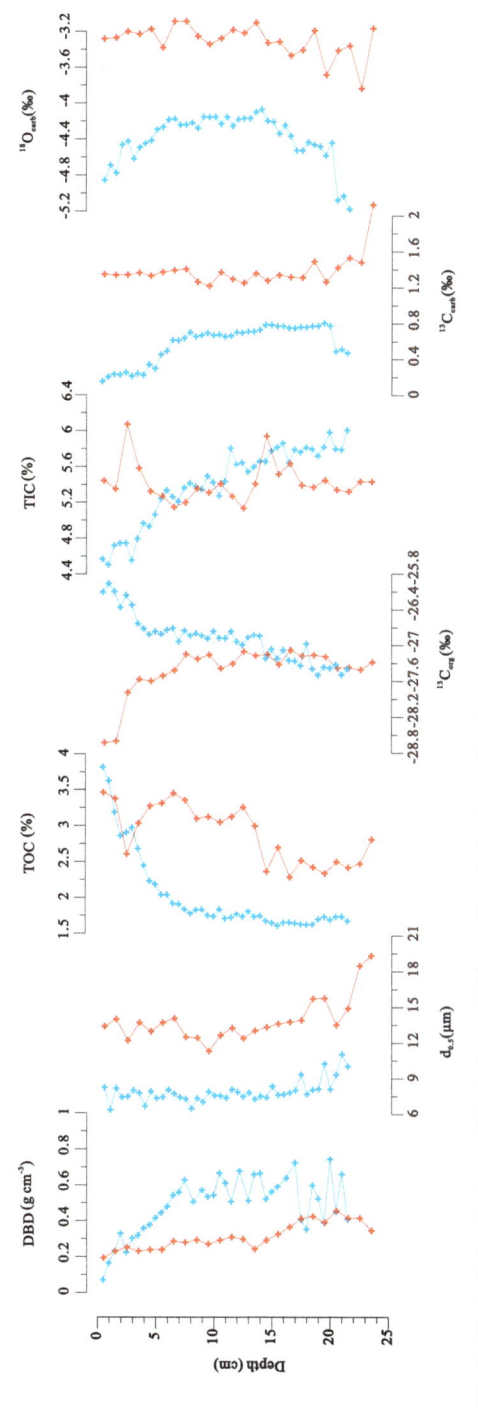

Fig. 5 Vertical distributions of the major variables at the B3 site (red colour) and B5 site (blue colour) (redrawn from Yu et al. 2015a)

Table 1 Properties of the mean, standard deviation (SD) and coefficient of variation (CV) in B3 core (after Yu et al. 2015a)

Sections (cm)		DBD (g cm^{-3})	d (0.5) (μm)	TOC (%)	$\delta^{13}C_{org}$ (‰)	TIC (%)	$\delta^{13}C_{carb}$ (‰)	$\delta^{18}O_{carb}$ (‰)
0.0–6.5	Mean	0.24	13.5	3.21	−27.88	5.46	1.35	−3.23
	SD	0.03	0.66	0.30	0.52	0.30	0.02	0.09
	CV	0.12	0.05	0.09	−0.02	0.06	0.02	−0.03
6.5–19.0	Mean	0.32	13.4	2.81	−27.2	5.42	1.32	−3.30
	SD	0.06	1.27	0.39	0.09	0.20	0.07	0.16
	CV	0.18	0.09	0.14	0	0.04	0.06	−0.05
19.5–23.5	Mean	0.41	16.6	2.55	−27.36	5.39	1.64	−3.44
	SD	0.05	2.78	0.18	0.05	0.06	0.33	0.28
	CV	0.11	0.17	0.07	0	0.01	0.20	−0.08
0.0–23.5	Mean	0.31	13.9	2.89	−27.43	5.42	1.38	−3.30
	SD	0.07	1.85	0.40	0.41	0.22	0.18	0.18
	CV	0.24	0.13	0.14	−0.01	0.04	0.13	−0.05

Table 2 Properties of the mean, standard deviation (SD) and coefficient of variation (CV) in B5 core (after Yu et al. 2015a)

Sections (cm)		DBD (g cm^{-3})	d (0.5) (μm)	TOC (%)	$\delta^{13}C_{org}$ (‰)	TIC (%)	$\delta^{13}C_{carb}$ (‰)	$\delta^{18}O_{carb}$ (‰)
0.0–6.0	Mean	0.31	7.59	2.75	−26.46	4.80	0.27	−4.50
	SD	0.12	0.58	0.59	0.32	0.25	0.10	0.21
	CV	0.39	0.08	0.22	−0.01	0.05	0.38	−0.05
6.0–18.5	Mean	0.57	7.67	1.74	−26.98	5.59	0.70	−4.25
	SD	0.11	0.51	0.09	0.20	0.20	0.05	0.13
	CV	0.20	0.07	0.05	−0.01	0.04	0.08	−0.03
18.5–21.5	Mean	0.52	9.47	1.71	−27.42	5.87	0.63	−4.80
	SD	0.14	1.18	0.03	0.07	0.12	0.16	0.33
	CV	0.27	0.12	0.02	0	0.02	0.26	−0.07
0.0–21.5	Mean	0.49	7.90	2.02	−26.9	5.41	0.57	−4.40
	SD	0.17	0.90	0.56	0.38	0.44	0.21	0.27
	CV	0.34	0.11	0.28	−0.01	0.08	0.37	−0.06

similar trend to TIC at both sites, showing little change at the B3 site but pronounced depletion of ^{13}C in the upper section at the B5 site. A large fluctuation was found in $\delta^{18}O_{carb}$ with a modest depletion below 12 cm at the B3 site. For the site B5, there were two regime shifts in $\delta^{18}O_{carb}$, i.e. significant enrichment of ^{18}O at around 14 cm and sharp depletion in the upper 6 cm. Interestingly, the average TIC content was the same for the two sites, whereas both $\delta^{13}C_{carb}$ and $\delta^{18}O_{carb}$ values were significantly different between the B3 site and the B5 site (Tables 1 and 2).

4 Temporal Variations of Carbon Burial Rates

Carbon burial rate (CBR, g C cm^{-2} year^{-1}) was determined as follows:

$$C_x BR = SR \times C_x \times 100 \qquad (5)$$

where C_x represented each carbon stock (TC, TOC or TIC).

The burial rate of TOC showed an almost line increase prior to 1990 at both sites, and there was a large similarity in the magnitude and temporal variability between the B3 and B5 sites (Fig. 6). For example, TOC burial rate varied from <10 g C m^{-2} year^{-1} in the 1950s to ~40 g C cm^{-2} year^{-1} post 2000 at both sites. However, there were large differences in the magnitude and temporal variability of TIC burial between the B3 and B5 sites. Overall, burial rate of TIC was much greater at the B5 site than at the B3 site, particularly over the period of 1960–2000, i.e. 25–90 g C m^{-2} year^{-1} versus 40–126 g C m^{-2} year^{-1}. There was a strong interannual variability in TIC burial at both sites, but the two peaks appeared in different years, i.e. around 1994 and 2008 at the B3 site, but 1985 and ~2000 at the B5 site. In addition, the two sites revealed opposite changing trends post 2000, i.e. an overall increasing trend at the B3 site but a sharp decreasing trend at the B5 site (from >100 g C m^{-2} year^{-1} down to ~50 g C m^{-2} year^{-1}). Because of the dominant of TIC burial, TC burial rate showed a similar interannual variability to TIC at both sites.

5 Underlying Mechanisms Responsible for the Temporal Variability of Carbon Burial

Since 1950, there has been a rapid increase in the deposition rate of carbon in many lakes at home and abroad, which means more carbon is fixed in lake sediments. This increasing trend in lake carbon burial is also recorded in our Bosten Lake. To determine which factors might be responsible for the temporal variation of carbon burial, we conducted correlation analyses using various variables in the sediments of B5 site (at the deepest location and away from the river mouths thus with little turbulence and resuspension) and other variables (such as lake level, air temperature, precipitation and evaporation). This study pointed out that climate change (i.e. temperature and evaporation) and human disturbance (i.e. land-use change and nutrient inputs) were the main driving forces over the past decades in the Bosten Lake.

5.1 Climatic Factor

There has a close and complex relationship between lake carbon burials and climate change. Generally speaking, the increase of rainfall will enhance the carbon storage

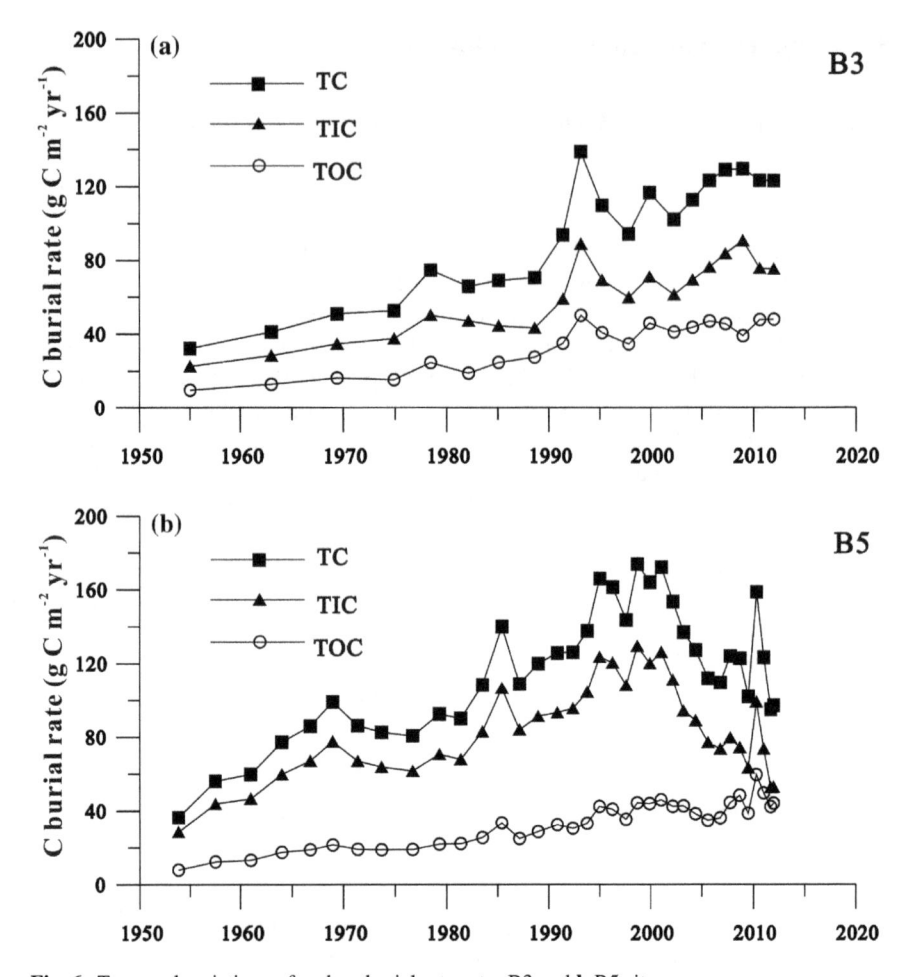

Fig. 6 Temporal variations of carbon burial rates at a B3 and b B5 sites

of soil and vegetation in the lake basin, and much more particulate organic carbon and dissolved organic carbon will be transported into the lake by the river run-off. In addition, increased rainfall will accelerate the weathering of carbonate in the basin, resulting in the increase of dissolved inorganic carbon in rivers and exogenous carbonate components in lakes. These findings have been confirmed in many lakes in China, such as Qinghai Lake (Xu et al. 2013), Daihai Lake (Xiao et al. 2004) and Dali Lake (Xiao et al. 2006), reflecting that increased rainfall can promote the lake's carbon sequestration. However, there was not significant relationship between carbon burial and rainfall in Bosten Lake. For one thing, the Bosten Lake is in arid region and the rainfall here is very little; therefore, the direct lake precipitation has less influence on the lake. For another, a variety of water conservancy and control facilities constructed by humans in the Kaidu River interfere with the effects of natural

Table 3 Correlation coefficient (r) between the main variables in the B5 core (after Yu et al. 2015a)

	TOC BR′	TIC BR	$^{13}C_{org}$	$^{13}C_{carb}'$	$^{18}O_{carb}$	Lake level	Evaporation	Precipitation
TIC BR	0.90[c]							
$^{13}C_{org}$	0.53[b]	0.17						
$^{13}C_{carb}'$	0.92[b]	0.75[c]	0.70[c]					
$^{18}O_{carb}$	0.06	0.41[a]	−0.50[b]	−0.01				
Lake level	−0.04	0.13	−0.46[a]	−0.23	0.16			
Evaporation	0	0.10	0	0.11	0.69[c]	−0.55[b]		
Precipitation	0.27	0.24	−0.09	0.18	0.09	0.03	0	
Air temperature	0.77[c]	0.61[c]	0.66[c]	0.82[c]	−0.09	−0.23	−0.04	0.05

TOC BR′ = initial burial rate of TOC, $^{13}C_{carb}'$ = the normalized $^{13}C_{carb}$. Significance of Pearson correlation is marked with [a]($p < 0.05$), [b]($p < 0.01$) and [c]($p < 0.001$) asterisks

rainfall on lakes. Furthermore, the relatively small time scale and low resolution may also be responsible for this discrepancy.

In general, an increase in temperature could lead to longer growing seasons especially in the boreal and temperate zones (Dong et al. 2012; Houghton 2007), which would promote biological productivity in the water column and enhance sedimentation of organic materials. However, there was also evidence that warming could stimulate mineralization of OC in sediments (Gudasz et al. 2010). Recent studies in China have shown that increased temperature promotes the primary productivity and enhances the carbon burial in lakes (Dong et al. 2012; Gui et al. 2013).

As illustrated in Fig. 5, TOC showed an obvious decrease over depth, with significantly lower values below 8 cm, which might primarily result from the decomposition of older sediments. To better understand the temporal variability of TOC burial, we have eliminated the effect of decomposition and then used the predicted initial TOC stock to calculate the initial TOC burial rate (Yu et al. 2015a). We found that there was a good correlation between the initial TOC burial rate and the annual mean air temperature (Table 3), indicating that the increased burial rate of TOC might be attributable to warming in the Bosten Lake. In addition, we found that TIC burial rate also had a good correlation with the annual mean air temperature ($r = 0.61$, $p < 0.001$). Higher temperature would lead to lower solubility of carbonate in lake water column, thus higher rate of carbonate precipitation (Leng and Marshall 2004).

In lakes, the carbonate production is mainly determined by the water chemistry (Kelts and Hsu 1978). Our analyses indicated that there was significant correlation between TIC burial rate and the initial burial rate of TOC in the Bosten Lake. Higher TIC burial rate might be associated with higher rate of photosynthetic utilization of CO_2 in the water column, indicating the nature of autochthonous carbonate (Paprocka 2007; Yu et al. 2015b). Chemical properties of lake water can change during the process of photosynthesis, resulting in carbonate supersaturation, and post-depositional diagenetic carbonates (Kelts and Hsu 1978; Li et al. 2012). Generally, an increase of lake productivity might lead to an enrichment of ^{13}C in carbonate (Leng et al. 2006; Zhu et al. 2013). However, $\delta^{13}C_{carb}$ in the Bosten Lake did not increase during 1950–2002 when TOC burial rate presented an increasing trend. On the one hand, some studies indicated that the effect of primary productivity on carbonate precipitation was not significant due to the infiltration of some terrestrial organic matter or the relatively low primary productivity in arid lakes (Xu et al. 2006). On the other hand, it was possible that the Suess effect could lead to more negative $\delta^{13}C_{carb}$ imprint in later years. Apparently, it is better to use normalized $\delta^{13}C_{carb}$ values (i.e. the Suess effect eliminated) to investigate the temporal variability. The normalized $\delta^{13}C_{carb}$ value for a certain year was calculated by adding 0.026‰ year^{-1}, according to the changed $\delta^{13}C$ values in the atmospheric CO_2 (Keeling et al. 2001). The normalized $\delta^{13}C_{carb}$ was highly correlated with the initial burial rate of TOC ($r = 0.92, P < 0.001$), indicating that carbonate formation was associated with the lake productivity in the Bosten Lake.

5.2 Anthropogenic Factor

In the past century, with the increase of the population in the river basin, human activities have become more and more important to the lakes and watershed ecosystems. Land-use changes, such as transforming primitive forest, grassland and desert ecosystems into farmland system, have greatly affected the vegetation type, land-use efficiency and crop area of the watershed by cutting, ploughing and discarding. On the one hand, the carbon stored in forest, grassland or soil is discharged directly into the atmosphere in CO_2 form. On the other hand, more terrestrial carbonaceous substances enter into the lake with river run-off.

Our analyses showed a decrease of air temperature during some periods (e.g. 1965–1968, 1973–1976, 1982–1984 and 1990–1992), but little decrease in TOC burial rate (Fig. 7). There was no relationship between the initial burial rate of TOC and air temperature ($r = 0.25, p = 0.18$), implying that other factors or processes might have affected the TOC burial in the Bosten Lake. The burial rate of TOC is influenced by water column productivity and external inputs. Autochthonous and allochthonous TOC have different isotopic signal. According to the analysis by Yu et al. (2015c), $^{13}C_{org}$ was more negative in autochthonous TOC than in allochthonous TOC in Bosten Lake. There might be an increase of allochthonous contribution, because $^{13}C_{org}$ presented an enrichment over time, from $-27.4‰$ prior to 1950 to $-26.5‰$ during 2002–2012. Indeed, studies showed large impacts of human activities on primary productivity of Bosten Lake over the twentieth century (Wünnemann et al. 2006; Zheng 2012).

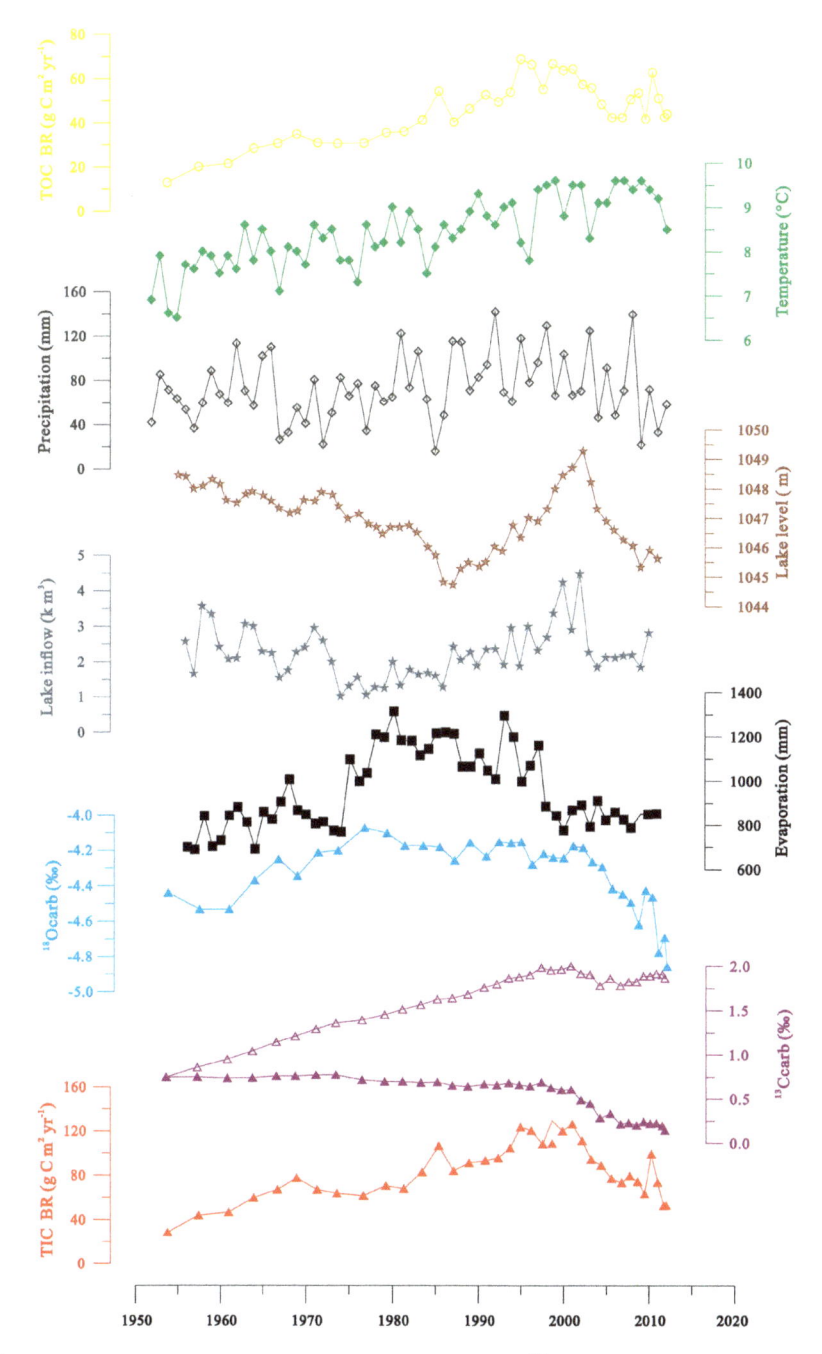

Fig. 7 Time series of the burial rates of initial TOC and TIC, $^{13}C_{carb}$ (solid triangles), normalized $^{13}C_{carb}$ (hollow triangles), $^{18}O_{carb}$ and meteorological (temperature, precipitation and evaporation) and hydrological data (lake level) (after Yu et al. 2015a)

Anthropogenic changes in land use and land cover (conversion of native lands to intensive agriculture lands) in the watershed could lead to enhanced carbon burial in lakes due to terrestrial inputs of organic and inorganic carbon via run-off and enhanced biological productivity as a results nutrient inputs (Anderson et al. 2013, 2014). The Yanqi Basin experienced a remarkable reclamation around 1970, followed by large population growth since 1980s (Zheng 2012), which could cause changes in the hydrological and/or geochemical processes in the watershed with consequences for the carbon cycle. In particular, agricultural practice (with irrigation and fertilization) could supply extra nutrients via run-off, thus enhance water column biological production and TOC burial in sediment of the Bosten Lake. Furthermore, the inlet rivers also brought inorganic detritus, minerals and salts into the lake, which were beneficial to the saturation of carbonate. Interestingly, TIC burial showed a sharp decrease during the period of 2002–2012 when temperature was generally high (Fig. 7). The decline of TIC burial post 2002 might result from less evaporation and the reduced biological activity (Zhang et al. 2009). In addition, lake level decline would cause a decrease in the total amount of carbonate in the water body then reduced the TIC burial rate in the sediment.

6 Conclusion

In this study, we evaluated multi-variables of two short sediment cores collected from the Bosten Lake and analysed the relationships between the burial rates of TOC and TIC and main climatic and hydrological variables. Our analyses indicated that carbon burials revealed a profound temporal variability with an overall increasing trend, and the TIC burial rate was proximately twice of the TOC since 1950. The increasing trend of TOC might be attributable to warming and increased human activities in the region. Warming also led to a significant increase in TIC burial during 1950–2002. Reduced biological activity, less evaporation and a rapid decline in lake level may have led to a sharp decrease in the TIC burial in the last 10 years (2002–2012). This study pointed out that carbonate burial in arid area might be an effective means for carbon sequestration, but TIC burial was much more sensitive than TOC in responding to climate change and human disturbance.

References

Alin SR, Johnson TC (2007) Carbon cycling in large lakes of the world: a synthesis of production, burial, and lake-atmosphere exchange estimates. Glob Biogeochem Cycles 21:GB3002. https://doi.org/10.1029/2006gb002881

Anderson NJ, Bennion H, Lotter AF (2014) Lake eutrophication and its implications for organic carbon sequestration in Europe. Glob Change Biol 20:2741–2751. https://doi.org/10.1111/gcb.12584

Anderson NJ, Dietz RD, Engstrom DR (2013) Land-use change, not climate, controls organic carbon burial in lakes. Proc R Soc B: Biol Sci 280:3907–3910. https://doi.org/10.1098/rspb.2013.1278

Appleby PG (2001) Chronostratigraphic techniques in recent sediments. In: Last W, Smol J (eds) Tracking environmental change using lake sediments. Springer, Netherlands

Battin TJ, Luyssaert S, Kaplan LA, Aufdenkampe AK, Richter A, Tranvik LJ (2009) The boundless carbon cycle. Nat Geosci 2:598–600

Blais JM, Kalff J (1995) The influence of lake morphometry on sediment focusing. Limnol Oceanogr 40:582–588

Chen F, Huang X, Zhang J, Holmes JA, Chen J (2006) Humid Little Ice Age in arid central Asia documented by Bosten Lake, Xinjiang, China. Sci China Ser D 49:1280–1290

Cole JJ, Prairie YT, Caraco NF, McDowell WH, Tranvik LJ, Striegl RG, Duarte CM, Kortelainen P, Downing JA, Middelburg JJ, Melack J (2007) Plumbing the global carbon cycle: integrating inland waters into the terrestrial carbon budget. Ecosystems 10:172–185. https://doi.org/10.1007/s10021-006-9013-8

Dean WE, Gorham E (1998) Magnitude and significance of carbon burial in lakes, reservoirs, and peatlands. Geology 26:535–538

Dong X, Anderson NJ, Yang X, Chen X, Shen J (2012) Carbon burial by shallow lakes on the Yangtze floodplain and its relevance to regional carbon sequestration. Glob Change Biol 18:2205–2217. https://doi.org/10.1111/j.1365-2486.2012.02697.x

Downing JA, Cole JJ, Middelburg JJ, Striegl RG, Duarte CM, Kortelainen P, Prairie YT, Laube KA (2008) Sediment organic carbon burial in agriculturally eutrophic impoundments over the last century. Global Biogeochem Cycles 22:GB1018. https://doi.org/10.1029/2006GB002854

Einsele G, Yan J, Hinderer M (2001) Atmospheric carbon burial in modern lake basins and its significance for the global carbon budget. Global Planet Change 30:167–195. https://doi.org/10.1016/S0921-8181(01)00105-9

Gudasz C, Bastviken D, Steger K, Premke K, Sobek S, Tranvik LJ (2010) Temperature-controlled organic carbon mineralization in lake sediments. Nature 466:478–481

Gui Z, Xue B, Yao S, Wei W, Yi S (2013) Organic carbon burial in lake sediments in the middle and lower reaches of the Yangtze River Basin, China. Hydrobiologia 710:143–156. https://doi.org/10.1007/s10750-012-1365-9

Guo M, Wu W, Zhou X, Chen Y, Li J (2015) Investigation of the dramatic changes in lake level of the Bosten Lake in northwestern China. Theoret Appl Climatol 119:341–351. https://doi.org/10.1007/s00704-014-1126-y

Heathcote AJ, Downing JA (2012) Impacts of eutrophication on carbon burial in freshwater lakes in an intensively agricultural landscape. Ecosystems 15:60–70

Herczeg AL, Smith AK, Dighton JC (2001) A 120 year record of changes in nitrogen and carbon cycling in Lake Alexandrina, South Australia: C:N, δ15N and δ13C in sediments. Appl Geochem 16:73–84. https://doi.org/10.1016/S0883-2927(00)00016-0

Houghton RA (2007) Balancing the Global Carbon Budget. Annu Rev Earth Planet Sci 35:313–347. https://doi.org/10.1146/annurev.earth.35.031306.140057

Kastowski M, Hinderer M, Vecsei A (2011) Long-term carbon burial in European lakes: analysis and estimate. Global Biogeochemical Cycles 25:GB3019

Keeling CD, Piper SC, Bacastow RB, Wahlen M, Whorf TP, Heimann M, Meijer HA (2001) Exchanges of atmospheric CO_2 and $^{13}CO_2$ with the terrestrial biosphere and oceans from 1978 to 2000. I. Global Aspects

Kelts K, Hsu KJ (1978) Freshwater carbonate sedimentation. In: Lerman A (ed) Lakes-chemistry, geology, physics. Springer, New York

Leng M, Lamb A, Heaton TE, Marshall J, Wolfe B, Jones M, Holmes J, Arrowsmith C (2006) Isotopes in lake sediments. In: Leng M (ed) Isotopes in palaeoenvironmental research. Springer, Netherlands

Leng MJ, Marshall JD (2004) Palaeoclimate interpretation of stable isotope data from lake sediment archives. Quatern Sci Rev 23:811–831. https://doi.org/10.1016/j.quascirev.2003.06.012

Li B, Chen Y, Shi X, Chen Z, Li W (2013) Temperature and precipitation changes in different environments in the arid region of northwest China. Theoret Appl Climatol 112:589–596

Li X, Liu W, Xu L (2012) Carbon isotopes in surface-sediment carbonates of modern Lake Qinghai (Qinghai–Tibet Plateau): implications for lake evolution in arid areas. Chem Geol 300–301:88–96. https://doi.org/10.1016/j.chemgeo.2012.01.010

O'Beirne MD, Strzok LJ, Werne JP, Johnson TC, Hecky RE (2015) Anthropogenic influences on the sedimentary geochemical record in western Lake Superior (1800–present). J Great Lakes Res 41:20–29. https://doi.org/10.1016/j.jglr.2014.11.005

Paprocka A (2007) Stable carbon and oxygen isotopes in recent sediments of Lake Wigry, NE Poland: implications for Lake Morphometry and environmental changes. In: Todd ED, Rolf TWS (eds) Terrestrial ecology. Elsevier, Amsterdam

Wünnemann B, Mischke S, Chen F (2006) A Holocene sedimentary record from Bosten Lake, China. Palaeogeogr Palaeoclimatol Palaeoecol 234:223–238. https://doi.org/10.1016/j.palaeo.2005.10.016

Wang R, Ernst G, Gao Q (2003) The recent change of water level in the Bosten Lake and analysis of its causes. J Glaciol Geocryol 25:60–64

Xiao J, Wu J, Si B, Liang W, Nakamura T, Liu B, Inouchi Y (2006) Holocene climate changes in the monsoon/arid transition reflected by carbon concentration in Daihai Lake of Inner Mongolia. Holocene 16:551–560

Xiao J, Xu Q, Nakamura T, Yang X, Liang W, Inouchi Y (2004) Holocene vegetation variation in the Daihai Lake region of north-central China: a direct indication of the Asian monsoon climatic history. Quatern Sci Rev 23:1669–1679. https://doi.org/10.1016/j.quascirev.2004.01.005

Xu H, Ai L, Tan L, An Z (2006) Stable isotopes in bulk carbonates and organic matter in recent sediments of Lake Qinghai and their climatic implications. Chem Geol 235:262–275. https://doi.org/10.1016/j.chemgeo.2006.07.005

Xu H, Lan J, Liu B, Sheng E, Yeager KM (2013) Modern carbon burial in Lake Qinghai, China. Appl Geochem 39:150–155. https://doi.org/10.1016/j.apgeochem.2013.04.004

Yu Z, Wang X, Zhao C, Lan H (2015a) Carbon burial in Bosten Lake over the past century: impacts of climate change and human activity. Chem Geol 419:132–141. https://doi.org/10.1016/j.chemgeo.2015.10.037

Yu Z, Wang X, Zhao C, Lan H (2015b) Spatial variations of carbonate and isotopes in the surface sediment of the Bosten Lake. J Lake Sci 27:250–257

Yu ZT, Wang XJ, Zhang EL, Zhao CY, Liu XQ (2015c) Spatial distribution and sources of organic carbon in the surface sediment of Bosten Lake, China. Biogeosciences 12:6605–6615. https://doi.org/10.5194/bg-12-6605-2015

Zhang C, Mischke S, Zheng M, Prokopenk A, Guo F, Feng Z (2009) Carbon and oxygen isotopic composition of surface-sediment carbonate in Bosten Lake (Xinjiang, China) and its controlling factors. Acta Geol Sin-Engl Ed 83:386–395

Zheng B (2012) Study of eco-environmental evolution of Bosten Lake during last 200 years based on Chironomid Larvae subfossil. Nanjing Institute of Geography & Limnology, Chinese Academy of Sciences, Nanjing, Jiangsu, P.R. China

Zhu Z, Ja Chen, Zeng Y (2013) Abnormal positive delta C-13 values of carbonate in Lake Caohai, southwest China, and their possible relation to lower temperature. Quatern Int 286:85–93

Carbon Sequestration in Arid Lands: A Mini Review

Xiujun Wang, Jiaping Wang, Huijin Shi and Yang Guo

1 Introduction

The global carbon cycle is regulated by organic and inorganic biogeochemical reactions in the atmosphere, biosphere, lithosphere, and pedosphere (Nordt et al. 2000). The fundamental understanding of the carbon cycle has to deal with the transformations of various carbon forms, fluxes between different pools, variability in these pools and/or fluxes, and underlying mechanisms. However, the studies of terrestrial carbon cycle have focused on the atmosphere–land CO_2 exchanges (with or without vegetations) and stocks of above- and below-ground organic carbon pools.

The soil organic carbon (SOC) pool is an important element in the terrestrial carbon cycle, acting as both carbon sinks and sources. It is well known that atmospheric CO_2 can be fixed and converted to organic carbon via photosynthesis, and a part of biologically fixed organic carbon is transferred into soil, leading to the accumulation of SOC. The decomposition of SOC produces CO_2 that can be released into the atmosphere partly or entirely depending on the environmental conditions. The strength of the carbon sinks and sources linked to SOC depends on many factors, particularly climatic and soil conditions, and agricultural practice (Ganuza and Almendros 2003; Guo et al. 2006; Ise and Moorcroft 2006; Jarecki and Lal 2005; Jimenez and Lal 2006; Kätterer et al. 1998; Reichstein et al. 2002). Climatic conditions, especially temperature and precipitation, may be responsible for the spatial variations in the magnitudes of soil carbon sequestration (Feng et al. 2006; Freibauer et al. 2004; Guo et al. 2006; Hontoria et al. 2005; Paustian et al. 1998). Favorite temperature and soil moisture can cause high rates of SOC decomposition thus low rates of SOC accumulation. On the other hand, dry conditions and high salt content may inhibit microbial activity.

X. Wang (✉) · H. Shi · Y. Guo
College of Global Change and Earth System Science, Beijing Normal University, Beijing 100875, China
e-mail: xwang@bnu.edu.cn

J. Wang
College of Agriculture, Shihezi University, Shihezi 832000, China

© Springer Nature Singapore Pte Ltd. 2018
X. Wang et al. (eds.), *Carbon Cycle in the Changing Arid Land of China*,
Springer Earth System Sciences, https://doi.org/10.1007/978-981-10-7022-8_10

Fig. 1 The main pathways of carbon cycle in the coupled atmosphere-land system in the arid and semi-arid regions. Diagram is modified according to Nordt et al. (2000)

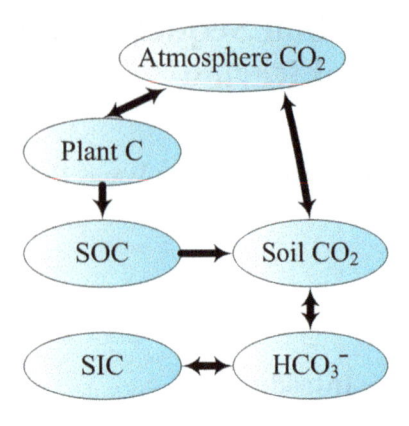

As an important carbon reservoir, SOC management has been put forward as one of the mitigating options for the global climate change (Lal 2002; Pan and Zhao 2005; Pan et al. 2008, 2009; Post et al. 2004). However, the pathway of carbon cycle also involves the dissolution of CO_2 and subsequent precipitation of soil carbonate or soil inorganic carbon (SIC), which often occurs in alkaline soils of arid and semi-arid regions (Fig. 1), but has not received much attention (Zamanian et al. 2016).

2 Precipitation of Soil Carbonate

The SIC pool consists of two major components: the lithogenic carbonate (LIC) and pedogenic carbonate (PIC). The former originates as detritus from parent materials, mainly limestone, whereas the latter (also termed as secondary carbonate) is formed by the dissolution and re-precipitation of LIC or through dissolution of CO_2 then precipitation with Ca^{2+} and/or Mg^{2+} from various minerals (e.g., silicate minerals, dust and chemical fertilizers). The large discrepancy in the estimated global SIC pool is probably due to the inability to differentiate fine primary carbonate in soils and carbonates of secondary origin (Eswaran et al. 2000).

The PIC formation via CO_2 dissolution in soils could lead to carbon sequestration (Monger and Gallegos 2000), which involves the following reactions:

$$CO_2 + H_2O \leftrightarrow H^+ + HCO_3^- \tag{1}$$

$$Ca^{2+} + 2HCO_3^- \leftrightarrow CaCO_3 + H_2O + CO_2 \tag{2}$$

In general, soils of humid and semi-humid regions are acidic, which drives the reaction (1) to the left, leading to CO_2 release into the atmosphere thus little SIC accumulated in soil profiles. However, soils in arid and semi-arid lands have high pH (often >8) and there are high levels of Ca^{2+} and/or Mg^{2+} in the groundwaters and river waters, which favor the precipitation of $CaCO_3$ in soil profiles.

An increase of CO_2 concentration and/or water in soil pores could also drive the reaction (1) to the right, i.e., the formation of HCO_3^-. A modeling study suggests that under elevated CO_2 concentration, there will be translocation of inorganic carbon from the upper layers to deep soils (Hirmas et al. 2010). Therefore, extra Ca^{2+} in soil profile would lead to $CaCO_3$ precipitation. Indeed, a recent study demonstrates that an increase of CO_2 concentration in saline soils can enhance $CaCO_3$ precipitation (Zhao et al. 2016b). There may be a few steps involved in the process of $CaCO_3$ precipitation in arid lands: (1) production of CO_2 due to SOM decomposition and root respiration in the upper horizons, which leads to formation of HCO_3^-, (2) transportation of HCO_3^- by water into the subsoil horizons, and (3) precipitation of $CaCO_3$ in the subsoil horizons (Lal and Kimble 2000).

The belowground biomass and carbon accumulation often exceed aboveground significantly in arid and semi-arid ecosystems (Fornara and Tilman 2008; Lufafa et al. 2009). Thus, one would expect that significant proportion of the absorbed carbon might be distributed below ground and stored in various carbon pools, including SOC and SIC pools. Indeed, an early report shows that the Saskatchewan soils in Canada have sequestrated 1.4 times more carbon in the form of PIC than as SOC (Landi et al. 2003).

3 Relationship Between SIC and SOC Stocks

There have been much less studies on the magnitudes and variability of SIC relative to SOC, and limited studies show inconsistent findings on the relationship between SIC and SOC. On the one hand, a few studies report a negative correlation between SIC and SOC stocks in the northern China (Li et al. 2010; Pan et al. 2000; Zhao et al. 2016a). On the other hand, some recent studies show that there is a significantly positive correlation between SIC and SOC stocks over the 0–100 cm layer in the croplands of arid and semi-arid regions, including the Yanqi Basin (Wang et al. 2015b), the upper Yellow River Delta (Guo et al. 2016) and the North China Plain (Shi et al. 2017). A positive correlation between SIC and SOC also exists in the soil profiles of various ecosystems in the Northern China (Gao et al. 2017; Zhang et al. 2010a).

The disagreement in the relationship between SIC and SOC might be associated with the differences in climate conditions, soil properties and management practices, which affect the processes of precipitation and dissolution of carbonate. Taking our recent analyses (Shi et al. 2017) as an example, there is a weak relationship between SIC and SOC in the topsoil ($R = 0.38$, $P = 0.69$), but a significantly positive correlation between SIC and SOC over the 0–100 cm ($R = 0.74$, $P < 0.001$) in the North China Plain; In addition, there are some differences in the relationship between SIC and SOC even under the same soil type and cropping system.

As shown in Fig. 2, the slope in the linear relationship between SIC and SOC is greater in the upper Yellow River Delta than in the North China Plain. Interestingly, the combined dataset show a strong positive correlation between SIC and SOC stocks

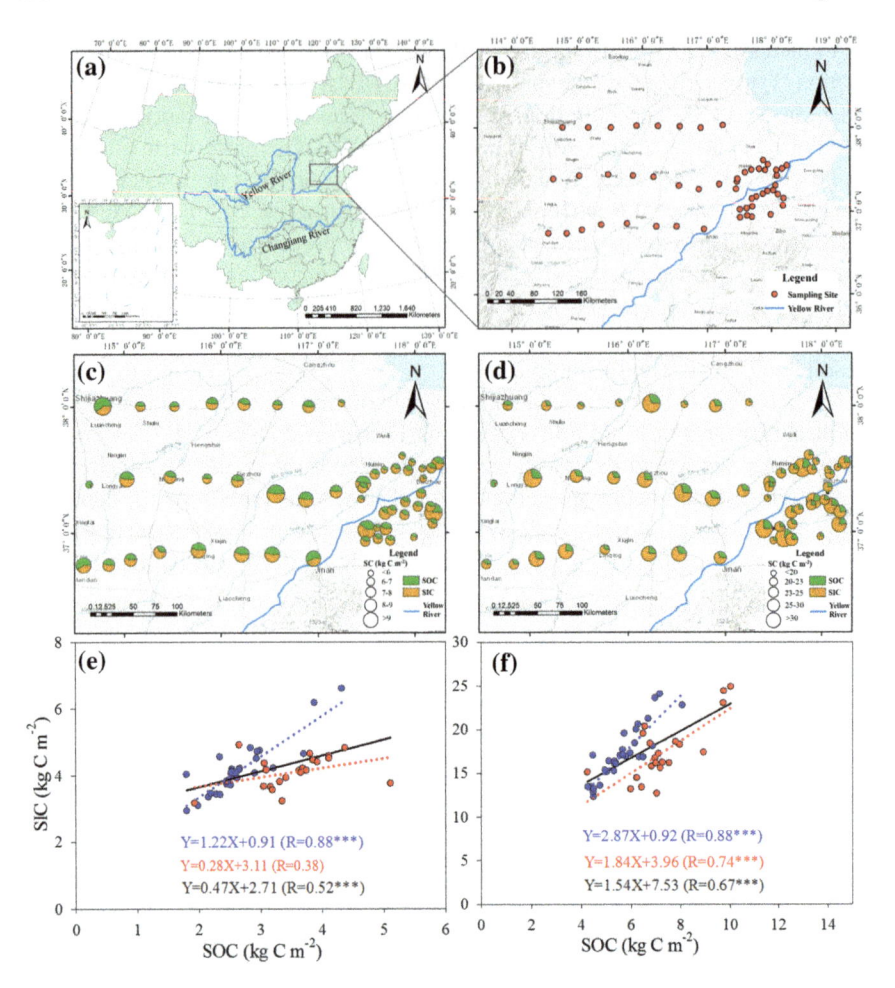

Fig. 2 The map of sampling sites (**a**, **b**), and spatial distributions of soil inorganic carbon (SIC) and organic carbon (SOC) stocks and their relationships for the 0–30 cm (**c**, **e**) and 0–100 cm (**d**, **f**). Blue circles and dotted line in **e** and **f** are for the upper Yellow River delta, red circles and dotted line for Hebei Plain, and black solid line for the combined data. Plots are redrawn from Shi et al. (2017). ***Significant at $P < 0.001$

over both the 0–30 cm layer ($R = 0.52$, $P < 0.001$) and 0–100 cm layer ($R = 0.67$, $P < 0.001$). However, some studies indicate that the relationship between SIC and SOC is negative in the surface soil of northern China, e.g. the cropland of Hebei Plain (Li et al. 2010), the cropland and other land uses of Chinese Loess Plateau (Zhao et al. 2016a). The findings from these limited studies indicate the relationship between SIC and SOC is complicated, which might reflect the decoupling of various processes involved with the formation of carbonate over time and space (Zheng et al. 2011).

While both negative and positive correlations may exist, water limitation and saline/alkaline conditions in the arid and semi-arid regions would be beneficial to the carbonate formation. Thus, a positive relationship would be more common in the cropland of northern China, in particular, when SOC and SIC stocks in deep soils are included, which implies that increasing SOC may lead to an increase of SIC in arid and semi-arid lands. Indeed, some studies on the cropland of northern China revealed that increasing SOC through long-term application of organic materials can significantly enhance carbonate accumulation (Bughio et al. 2016; Wang et al. 2014; Wang et al. 2015b).

4 Impacts of Land Use Changes

Land use and management practice have large influences on the biogeochemical processes in the soil-plant systems in many aspects. Good land management can improve soil physical and chemical conditions, and promote microbial and biological activities, which has implications for the carbon and nutrient cycling. There are many studies addressing the impacts of land use changes on soil carbon dynamics, which yield inconsistent findings. On the one hand, there is evidence that tillage during farming can cause a decline in SOC in some areas, particularly in the temperate and tropical regions (Conant et al. 2001; Murty et al. 2002; Ogle et al. 2005). However, some studies have showed that there is an increase in SOC of the topsoil in the cropland (relative to native lands) of arid and semi-arid regions in the northwest China (Fan et al. 2008; Zhang et al. 2010b). The former (i.e., the decrease in SOC) is largely due to enhanced decomposition as a results of tillage whereas the latter (i.e., the increase of SOC) is owing to higher above- and below-ground biomass that is resulted from fertilization and irrigation (Wang et al. 2015b).

Limited studies have also showed that cropping can lead to a greater increase in SIC than other types of land use in arid and semi-arid regions, e.g., in the middle of Hexi Corridor, China (Su et al. 2010) and in the Russian Chernozem (Mikhailova and Post 2006). In addition, a number of studies have demonstrated that both SOC and SIC stocks are much higher in the cropland than in other types of land uses (e.g., grassland, desert, shrub land) in the northern China (Liu et al. 2014; Su et al. 2010; Tan et al. 2014; Wang et al. 2015a). The study conducted in the Yanqi Basin (Wang et al. 2015b) showed that the conversion of shrub land to cropland led to a significant increase of SOC in the topsoil and an increase of SIC in the subsoil. The increase of surface SOC in arid cropland is a result of intensive cropping (through fertilization and irrigation) that promotes plant growth and enhances organic carbon inputs into the topsoil (Khan et al. 2009; Minasny et al. 2012; Turner et al. 2011). The increase of SIC may be linked to increased CO_2 production in soil profile (Lopez-Sangil et al. 2013), as a result of an increase in both SOC decomposition (due to increased SOC) and root respiration (due to enhanced plant growth) (Wang et al. 2015b).

5 Implications and Future Directions

Arid and semi-arid regions cover more than 35% of Earth's land surface. There has been evidence of significant carbon sequestration in various ecosystems, including in deserts, agricultural lands, groundwaters and inland water bodies in arid and semi-arid regions. For example, the study of Xie et al. (2009) revealed significant CO_2 absorption in the alkaline soils of northwest China, which were influenced by soil temperature, water content, salinity and alkalinity. Another study (Wohlfahrt et al. 2008) also showed large annual net ecosystem CO_2 uptake by the Mojave Desert, USA. Limited analyses suggest that a large amount of absorbed CO_2 may be stored in the groundwater below the deserts (Li et al. 2015; Monger et al. 2015). The integrated studies carried out in the Yanqi Basin and Bosten Lake demonstrate that there has been significant accumulation of carbonate in soils and sediments, highlighting the dynamic nature of carbonate and the potential of arid lands for carbon storage. However, there are limited studies on the processes of carbonate formation, particularly the formation of pedogenic carbonate (Zamanian et al. 2016), and there is a need to improve our understanding how environmental changes affect the formation and transformation of carbonate (Gallagher and Sheldon 2016).

The arid and semi-arid regions have experienced significant climate changes and human activity over the past decades. For instance, there is a significant increase in precipitation in northwest China since 1950s (Wang and Zhou 2005), with implications for the terrestrial ecosystem and carbon cycle (Liu et al. 2009). There has been evidence that irrigation can enhance the accumulation of carbonate in arid and semi-arid lands (Bughio et al. 2016; Wang et al. 2016). Intensive cropping has resulted in a significant increase in SIC levels in the soils of northwest China (Wang et al. 2015a; Wu et al. 2009). With the increasing population and food demand, more native lands may be converted to agricultural lands in arid and semi-arid regions, which could lead to more carbon storage with sound agricultural practices. Apparently, more studies are needed at various scales to better understand the impacts of human activity and climate changes on carbon sequestration in the vast arid and semi-arid regions.

References

Bughio MA, Wang P, Meng F, Qing C, Kuzyakov Y, Wang X, Junejo SA (2016) Neoformation of pedogenic carbonates by irrigation and fertilization and their contribution to carbon sequestration in soil. Geoderma 262:12–19. https://doi.org/10.1016/j.geoderma.2015.08.003

Conant RT, Paustian K, Elliott ET (2001) Grassland management and conversion into grassland: effects on soil carbon. Ecol Appl 11:343–355

Eswaran H, Reich PF, Kimble JM, Beinroth FH, Padmanabhan E, Moncharoen P (2000) Global carbon stocks. In: Lal R, Kimble JM, Eswaran H, Stewart BA (eds) Global climate change and pedogenic carbonates. Lewis Publishers, Boca Raton, FL

Fan TL, Xu MG, Song SY, Zhou GY, Ding LP (2008) Trends in grain yields and soil organic C in a long-term fertilization experiment in the China Loess Plateau. J Plant Nutr Soil Sci 171:448–457. https://doi.org/10.1002/jpln.200625192

Feng Q, Wei L, Zhang YW, Si JH, Su YH, Qiang CZ, Xi HY (2006) Effect of climatic changes and human activity on soil carbon in desertified regions of China. Tellus Ser B-Chem Phys Meteorol 58:117–128

Fornara D, Tilman D (2008) Plant functional composition influences rates of soil carbon and nitrogen accumulation. J Ecol 96:314–322

Freibauer A, Rounsevell MDA, Smith P, Verhagen J (2004) Carbon sequestration in the agricultural soils of Europe. Geoderma 122:1–23

Gallagher TM, Sheldon ND (2016) Combining soil water balance and clumped isotopes to understand the nature and timing of pedogenic carbonate formation. Chem Geol 435:79–91. https://doi.org/10.1016/j.chemgeo.2016.04.023

Ganuza A, Almendros G (2003) Organic carbon storage in soils of the Basque Country (Spain): the effect of climate, vegetation type and edaphic variables. Biol Fertil Soils 37:154–162. https://doi.org/10.1007/s00374-003-0579-4

Gao Y, Tian J, Pang Y, Liu JB (2017) Soil inorganic carbon sequestration following afforestation is probably induced by pedogenic carbonate formation in Northwest China. Front Plant Sci 8:1282. https://doi.org/10.3389/fpls.2017.01282

Guo YY, Gong P, Amundson R, Yu Q (2006) Analysis of factors controlling soil carbon in the conterminous United States. Soil Sci Soc Am J 70:601–612. https://doi.org/10.2136/sssaj2005.0163

Guo Y, Wang X, Li X, Wang J, Xu M, Li D (2016) Dynamics of soil organic and inorganic carbon in the cropland of upper Yellow River Delta, China. Sci Rep 6:36105. https://doi.org/10.1038/srep36105

Hirmas DR, Amrhein C, Graham RC (2010) Spatial and process-based modeling of soil inorganic carbon storage in an arid piedmont. Geoderma 154:486–494. https://doi.org/10.1016/j.geoderma.2009.05.005

Hontoria C, Rodriguez-Murillo JC, Saa A (2005) Soil organic carbon contents in arid and semiarid regions of peninsular Spain. Sustain Use Manage Soils—Arid Semiarid Reg 36:275–280

Ise T, Moorcroft PR (2006) The global-scale temperature and moisture dependencies of soil organic carbon decomposition: an analysis using a mechanistic decomposition model. Biogeochemistry 80:217–231. https://doi.org/10.1007/s10533-006-9019-5

Jarecki MK, Lal R (2005) Soil organic carbon sequestration rates in two long-term no-till experiments in Ohio. Soil Sci 170:280–291. https://doi.org/10.1097/01.ss.0000162286.95137.70

Jimenez JJ, Lal R (2006) Mechanisms of C sequestration in soils of Latin America. Crit Rev Plant Sci 25:337–365. https://doi.org/10.1080/0735268060094240

Kätterer T, Reichstein M, Andren O, Lomander A (1998) Temperature dependence of organic matter decomposition: a critical review using literature data analyzed with different models. Biol Fert Soils 27:258–262

Khan S, Hanjra MA, Mu J (2009) Water management and crop production for food security in China: a review. Agric Water Manag 96:349–360. https://doi.org/10.1016/j.agwat.2008.09.022

Lal R (2002) Soil carbon sequestration in China through agricultural intensification, and restoration of degraded and desertified ecosystems. Land Degrad Dev 13:469–478. https://doi.org/10.1002/ldr.531

Lal R, Kimble JM (2000) Pedogenic carbonate and the global carbon cycle. In: Lal R, Kimble JM, Eswaran H, Stewart BA (eds) Global climate change and pedogenic carbonate. CRC Press, Boca Raton, FL, USA

Landi A, Mermut AR, Anderson DW (2003) Origin and rate of pedogenic carbonate accumulation in Saskatchewan soils, Canada. Geoderma 117:143–156

Li GT, Zhang CL, Zhang HJ (2010) Soil inorganic carbon pool changed in long-term fertilization experiments in north China plain. World Cong Soil Sci Soil Solut Changing World 19:220–223

Li Y, Wang YG, Houghton RA, Tang LS (2015) Hidden carbon sink beneath desert. Geophys Res Lett 42:5880–5887. https://doi.org/10.1002/2015gl064222

Liu WX, Zhang Z, Wan SQ (2009) Predominant role of water in regulating soil and microbial respiration and their responses to climate change in a semiarid grassland. Glob Change Biol 15:184–195

Liu WG, Wei J, Cheng JM, Li WJ (2014) Profile distribution of soil inorganic carbon along a chronosequence of grassland restoration on a 22-year scale in the Chinese Loess Plateau. CATENA 121:321–329. https://doi.org/10.1016/j.catena.2014.05.019

Lopez-Sangil L, Rovira P, Casals P (2013) Decay and vertical reallocation of organic C, and its incorporation into carbonates, in agricultural soil horizons at two different depths and rewetting frequencies. Soil Biol Biochem 61:33–44. https://doi.org/10.1016/j.soilbio.2013.02.008

Lufafa A, Diedhiou I, Ndiaye NAS, Sene M, Kizito F, Dick RP, Noller JS (2009) Allometric relationships and peak-season community biomass stocks of native shrubs in Senegal's Peanut Basin. J Arid Environ 73:260–266

Mikhailova EA, Post CJ (2006) Effects of land use on soil inorganic carbon stocks in the Russian Chernozem. J Environ Qual 35:1384–1388. https://doi.org/10.2134/jeq2005.0151

Minasny B, McBratney AB, Hong SY, Sulaeman Y, Kim MS, Zhang YS, Kim YH, Han KH (2012) Continuous rice cropping has been sequestering carbon in soils in Java and South Korea for the past 30 years. Glob Biogeochem Cycles 26:GB3027. https://doi.org/10.1029/2012gb004406

Monger HC, Gallegos RA (2000) Biotic and abiotic processes and rates of pedogenic carbonate accumulation in the southwestern United States-Relationship to atmospheric CO_2 sequestration. In: Lal R, Kimble JM, Eswaran H, Stewart BA (eds) Golbal Climate Change and Pedogenic. CRC Press, Boca Raton, FL, USA

Monger HC, Kraimer RA, Khresat S, Cole DR, Wang X, Wang J (2015) Sequestration of inorganic carbon in soil and groundwater. Geology 43:375–378. https://doi.org/10.1130/g36449.1

Murty D, Kirschbaum MUF, McMurtrie RE, McGilvray H (2002) Does conversion of forest to agricultural land change soil carbon and nitrogen? A review of the literature. Glob Change Biol 8:105–123. https://doi.org/10.1046/j.1354-1013.2001.00459.x

Nordt LC, Wilding LP, Drees LR (2000) Pedogenic carbonate transformations in leaching soil systems: implications for the global C cycle. In: Lal R, Kimble JM, Eswaran H, Stewart BA (eds) Global climate change and pedogenic carbonates. CRC Press, Boca Raton, USA

Ogle SM, Breidt FJ, Paustian K (2005) Agricultural management impacts on soil organic carbon storage under moist and dry climatic conditions of temperate and tropical regions. Biogeochemistry 72:87–121. https://doi.org/10.1007/s10533-004-0360-2

Pan G, Zhao Q (2005) Study on evolution of organical carbon stock in agricultural soils of China: facing the challenge of global change and food security. Adv Earth Sci (in Chinese) 20:384–393

Pan GX, Guo T, Lal R, Kimble JM, Eswaran H, Stewart BA (2000) Pedogenic carbonate of aridic soils in China and its significance in carbon sequestration in terrestrial systems. In: Lal R, Kimble JM, Eswaran H, Stewart BA (eds) Global climate change and pedogenic carbonates. Lewis Publishers, Boca Raton

Pan GX, Wu LS, Li LQ, Zhang XH, Gong W, Wood Y (2008) Organic carbon stratification and size distribution of three typical paddy soils from Taihu Lake region, China. J Environ Sci-China 20:456–463

Pan G, Smith P, Pan W (2009) The role of soil organic matter in maintaining the productivity and yield stability of cereals in China. Agr Ecosyst Environ 129:344–348

Paustian K, Cole CV, Sauerbeck D, Sampson N (1998) CO2 mitigation by agriculture: an overview. Clim Change 40:135–162

Post WM, Izaurralde RC, Jastrow JD, McCarl BA, Amonette JE, Bailey VL, Jardine PM, West TO, Zhou JZ (2004) Enhancement of carbon sequestration in US soils. Bioscience 54:895–908

Reichstein M, Tenhunen JD, Roupsard O, Ourcival JM, Rambal S, Miglietta F, Peressotti A, Pecchiari M, Tirone G, Valentini R (2002) Severe drought effects on ecosystem CO2 and H2O fluxes at three Mediterranean evergreen sites: revision of current hypotheses? Glob Change Biol 8:999–1017

Shi HJ, Wang XJ, Zhao YJ, Xu MG, Li DW, Guo Y (2017) Relationship between soil inorganic carbon and organic carbon in the wheat-maize cropland of the North China Plain. Plant Soil 418:423–436. https://doi.org/10.1007/s11104-017-3310-1

Su YZ, Wang XF, Yang R, Lee J (2010) Effects of sandy desertified land rehabilitation on soil carbon sequestration and aggregation in an arid region in China. J Environ Manage 91:2109–2116. https://doi.org/10.1016/j.jenvman.2009.12.014

Tan WF, Zhang R, Cao H, Huang CQ, Yang QK, Wang MH, Koopal LK (2014) Soil inorganic carbon stock under different soil types and land uses on the Loess Plateau region of China. Catena 121:22–30. https://doi.org/10.1016/j.catena.2014.04.014

Turner NC, Molyneux N, Yang S, Xiong Y-C, Siddique KHM (2011) Climate change in south-west Australia and north-west China: challenges and opportunities for crop production. Crop Pasture Sci 62:445–456. https://doi.org/10.1071/CP10372

Wang Y, Zhou L (2005) Observed trends in extreme precipitation events in China during 1961–2001 and the associated changes in large-scale circulation. Geophysical Res Lett, 32:L09707. https://doi.org/10.1029/2005gl023769

Wang XJ, Xu MG, Wang JP, Zhang WJ, Yang XY, Huang SM, Liu H (2014) Fertilization enhancing carbon sequestration as carbonate in arid cropland: assessments of long-term experiments in northern China. Plant Soil 380:89–100. https://doi.org/10.1007/s11104-014-2077-x

Wang JP, Wang XJ, Zhang J, Zhao CY (2015a) Soil organic and inorganic carbon and stable carbon isotopes in the Yanqi Basin of northwestern China. Eur J Soil Sci 66:95–103. https://doi.org/10.1111/ejss.12188

Wang XJ, Wang JP, Xu MG, Zhang WJ, Fan TL, Zhang J (2015b) Carbon accumulation in arid croplands of northwest China: pedogenic carbonate exceeding organic carbon. Sci Rep 5:11439. https://doi.org/10.1038/srep11439

Wang JP, Monger C, Wang XJ, Serena M, Leinauer B (2016) Carbon Sequestration in Response to Grassland-Shrubland-Turfgrass Conversions and a Test for Carbonate Biomineralization in Desert Soils, New Mexico, USA. Soil Sci Soc Am J 80:1591–1603. https://doi.org/10.2136/sssaj2016.03.0061

Wohlfahrt G, Fenstermaker LF, Arnone Iii JA (2008) Large annual net ecosystem CO_2 uptake of a Mojave Desert ecosystem. Glob Change Biol 14:1475–1487. https://doi.org/10.1111/j.1365-2486.2008.01593.x

Wu HB, Guo ZT, Gao Q, Peng CH (2009) Distribution of soil inorganic carbon storage and its changes due to agricultural land use activity in China. Agr Ecosyst Environ 129:413–421. https://doi.org/10.1016/j.agee.2008.10.020

Xie JX, Li Y, Zhai CX, Li CH, Lan ZD (2009) CO_2 absorption by alkaline soils and its implication to the global carbon cycle. Environ Geol 56:953–961. https://doi.org/10.1007/s00254-008-1197-0

Zamanian K, Pustovoytov K, Kuzyakov Y (2016) Pedogenic carbonates: forms and formation processes. Earth Sci Rev 157:1–17. https://doi.org/10.1016/j.earscirev.2016.03.003

Zhang N, He XD, Gao YB, Li YH, Wang HT, Ma D, Zhang R, Yang S (2010a) Pedogenic carbonate and soil dehydrogenase activity in response to soil organic matter in Artemisia ordosica community. Pedosphere 20:229–235. https://doi.org/10.1016/s1002-0160(10)60010-0

Zhang WJ, Wang XJ, Xu MG, Huang SM, Liu H, Peng C (2010b) Soil organic carbon dynamics under long-term fertilizations in arable land of northern China. Biogeosciences 7:409–425

Zhao W, Zhang R, Huang CQ, Wang BQ, Cao H, Koopal LK, Tan WF (2016a) Effect of different vegetation cover on the vertical distribution of soil organic and inorganic carbon in the Zhifanggou Watershed on the loess plateau. Catena 139:191–198. https://doi.org/10.1016/j.catena.2016.01.003

Zhao X, Zhao C, Wang J, Stahr K, Kuzyakov Y (2016b) $CaCO_3$ recrystallization in saline and alkaline soils. Geoderma 282:1–8. https://doi.org/10.1016/j.geoderma.2016.07.004

Zheng J, Cheng K, Pan G, Pete S, Li L, Zhang X, Zheng J, Han X, Du Y (2011) Perspectives on studies on soil carbon stocks and the carbon sequestration potential of China. Chinese Science Bulletin 56:3748–3758. https://doi.org/10.1007/s11434-011-4693-7

Lightning Source UK Ltd.
Milton Keynes UK
UKHW02n1139070318
319042UK00001B/35/P